BERKELEY'S PHILOSOPHY OF SCIENCE

ARCHIVES INTERNATIONALES D'HISTOIRE DES IDEES

INTERNATIONAL ARCHIVES OF THE HISTORY OF IDEAS

65

RICHARD J. BROOK

BERKELEY'S PHILOSOPHY
OF SCIENCE

BERKELEY'S PHILOSOPHY OF SCIENCE

by

RICHARD J. BROOK

MARTINUS NIJHOFF / THE HAGUE / 1973

PRINTED IN THE NETHERLANDS

TABLE OF CONTENTS

INTRODUCTION

> Philonous: You see, Hylas, the water of yonder fountain, how it is forced upwards, in a round column, to a certain height, at which it breaks and falls back into the basin from whence it rose, its ascent as well as descent proceeding from the same uniform law or principle of gravitation. Just so, the same principles which at first view, lead to skepticism, pursued to a certain point, bring men back to common sense.[1]

Although major works on Berkeley have considered his Philosophy of

[1] George Berkeley, *Three Dialogues Between Hylas and Philonous,* ed. Colin Murray Turbayne, (third and final edition; London 1734); (New York: The Bobbs Merrill Company, Inc., Library of Liberal Arts, 1965), p. 211.
Berkeley, in general, conveniently numbered sections in his works, and in the text of the essay, we will refer if possible to the title and section number. References to the Three Dialogues Between *Hylas* and *Philonous* will be also made in the text and refer to the dialogue number and page in the Turbayne edition cited above.
The following editions of Berkeley's works have been used:
A Treatise Concerning the Principles of Human Knowledge, Three Dialogues Between Hylas and Philonous, and *Correspondence with Samuel Johnson,* ed. Colin Murray Turbayne, (Final 1734 edition of the first two works, four letters between Berkeley and Johnson first published as a unit in, *Samuel Johnson, President of King's College: His career and writings,* ed. Herbert and Avrol Schneider, (4 vols.; New York: Columbia University Press, 1929), 11, 261-284.)) (New York: Bobbs Merrill Co., Inc., 1965.)
An Essay Towards a New Theory of Vision (in the text called *"Essay"*).
The Theory of Vision, or *Visual Language Vindicated and Explained* (in the text called "Visual Language") ed. Colin Murray Turbayne, (*Essay* originally published in 1709, text here is Berkeley's fourth and last edition of 1732; Turbayne includes appendix of the second edition. Text of *Visual Language* from the 1733 (and only) edition.) (New York: The Bobbs Merrill Co., Inc.) (Library of Liberal Arts)) 1963.
De Motu, The Analyst, A Defense of Free-Thinking in Mathematics, Reasons for not Replying to Mr. Walton's Full Answer, Of Infinites, Ed. A. A. Luce (including Luce's English translation of *De Motu*), (Vol. 4 in *The Works of George Berkeley Bishop of Cloyne,* Ed. A. A. Luce and T. E. Jessop, (Thomas Nelson and Sons: London, 1951).
The Siris, Ed. T. E. Jessop, (Vol. 5 of *The Works of George Berkeley*) *op. cit.*
Philosophical Commentaries Ed. A. A. Luce (Vol. 1 of The Works of George Berkeley) *op. cit.*
Alciphron, or *The Minute Philosopher* Ed. T. E. Jessop (Vol. 3 of The Works of George Berkeley) *op. cit.*

Science, and important monographs have dealt with particular aspects of his scientific work, there appears to be no systematic and critical account of this work considered as a whole.[2] By "scientific work" we understand Berkeley's major writings in optics, physics, and mathematics as well as the passages in the more epistemological and metaphysical work that deal with methodological and substantive matters in science.[3] By a "critical account" we understand one that does more than assess Berkeley's historical place concerning problems in the Philosophy of Science, but seeks as well to assess the consistency and adequacy of his views considered in themselves.

Historical considerations are, of course, important. It can, and has been argued that Berkeley's views on mechanics are part of a long kinematical tradition that eschews the use of "forces" in physics and seeks explanations in terms of functional relations among such factors as changes of momenta, masses and distances.[4] It can be maintained, for example, that with Berkeley, as later with Ernst Mach, there is no problem with action at a distance. More generally, Berkeley consistently articulates a view widely held by the philosopher-scientists of the seventeenth and eighteenth century that there is no efficient causality within nature. From this perspective the alleged communication of motion through contact is as much a mystery as action at a distance, which is to say, it may be no mystery at all if one gives up the demand for "forces" as explanations.

In the Berkelian idom, this lack of efficient causality in nature, is

---

[2] Among the important monographs there are:
Gerard Hinrich, "The Logical Positivism of De Motu," *Review of Metaphysics,* III, 1950, 491-505.
W. A. Suchting, "Berkeley's Criticism of Newton on Space and Motion," *Isis,* Vol. 58, Summer 1967, 186-197.
G. W. R. Ardley, "Berkeley's Philosophy of Nature," Bulletin No. 63, Series No. 3, University of Auckland, 1962.
Karl Popper, "A Note on Berkeley as Precursor of Mach and Einstein," *British Journal for the Philosophy of Science,* 4, 1953; Reprinted, C. B. Martin and D. M. Armstrong ed; *Locke and Berkeley, A Collection of Critical Essays,* (New York: Doubleday and Co., Inc. (Anchor Books)), 1968, 436-449.
A Particularly good discussion of Berkeley's Philosophy of Science is found in: Gerd Buchdahl, *Metaphysics and the Philosophy of Science; The Classical Origins: Descartes to Kant.* Chapter V; "Berkeley: New Conceptions of Scientific Law and Explanation," (Cambridge: The MIT Press; 1969) 275-325.
A good discussion on Berkeley and the Newtonian concept of absolute space, stressing theological issues, is found in: Alexandre Koyre, *From the Closed World to the Infinite Universe,* Ch. X; "Absolute Space and Absolute Time; God's Frame of Action," (Berkeley and Newton) 221-235.
[3] See footnote 1.
[4] The view expressed by Max Jammer, *Concepts of Force; A Study in the Foundations of Dynamics,* (Cambridge: Harvard University Press, 1957) 203-208, 228.

expressed as "the passivity of ideas"; as with Leibniz it is expressed as the absence of causal interaction among monads.[5]

Berkeley was especially desirous of putting contemporary physics and mathematics in their proper perspective; stripping these sciences of pretensions they may have had to disclose the ultimate causal structure of reality. His strictures, therefore, often have an admittedly polemical cast; admonishing those who would criticize the doctrines of faith, to first put their own house in order.[6]

Nevertheless, the doctrine of the "passivity of ideas" strongly expresses the view that there is no efficient causality within nature, since "nature" viewed ultimately as the set of perceptual contents ("ideas") lacks within itself any reason for its existence. The proper object of science, then, is not to disclose efficient causes – the latter is properly the object of metaphysics or theology) – but rather to disclose the uniformities among the phenomena of nature. Like Leibniz, Berkeley draws a sharp distinction between metaphysics which deals with efficient, or moving (or "productive") causes and physics which does not. This distinction, while admittedly having as a fundamental intention the protection of religion, helps to clarify the nature of physics itself. When we consider *De Motu*, Berkeley's work in mechanics, we will be concerned with his own attempt to clarify the foundations of this science; particularly with his contention that "force" as it is used in mechanics, cannot in any sense be construed as a principle of animation.

We will not spend much time in criticism of the metaphysical doctrine of "immaterialism" although its demonstration is of crucial importance to Berkeley. He argues, for example, that his theory of vision offers evidence for the ideality of space; hence evidence for immaterialism. We consider his argument here to be mistaken; but the theory of vision does raise important conceptual issues. For example, the notion of "mathematical fictions" useful in science (in this case optics) which refer to nothing in reality.

The theory of vision raises, as well, the question of what is the object of geometry; a question pursued in the *Principles* and the *Analyst* and a question which in turn raises the more fundamental problem of how we

---

[5] The "passivity" of matter is in fact a general theme of 16th and 17th century philosophy reflecting its break with Aristotelianism. As we will see, however, even after rejecting "innate" or animating principles in matter, we are still faced with the question of the ontological status of "impressed forces."

[6] This is particularly true in the *Analyst*, where the polemical cast of the work is quite explicit. For example: "But he who can digest a second or third fluxion, a second or third difference, need not me thinks, be squeamish about any point in divinity." (*Analyst* 7)

are to understand the application of mathematics to the physical world. Berkeley strongly opposed that tradition which viewed Euclidean Geometry as the science not of material bodies, but of pure space, of *extension simpliciter*.[7] Our concern, however, will not be as much with the issue of whether space is transcendentally real; that is an existent, independent of perception and a container of physical processes; but with the methodological problem of distinguishing "pure" from "applied" geometry. Is geometry for Berkeley, a straight forwardly empirical discipline dealing with certain characteristics of experienced objects, or is it a formal system whose fundamental terms have no extra systematic reference?

The claim that geometry articulates the properties of "sensible" extension (extension apprehended visually or tactually) is related to Berkeley's important contention that finite perceptually apprehended extension is composed of a finite number of "sensible minima." Although Berkeley considered this contention to follow in important ways from the truth of "immaterialism,"[8] we will focus on the doctrine in the context of problems in the concept of the actual infinite, or in Berkeley's words problems with the "infinite divisibility" of finite extension.[9]

Since Berkeley's Philosophy of Science, construed as clarification of the foundations of science, often takes the form of an analysis of the meaning of scientific and mathematical language, we will first consider his theory of the significance or meaning of language in general.

After considering Berkeley's general theory of signification, (meaning) we will consider in some detail his theory of "signs." The latter topic is an essential part (not the whole) of the former; for in nothing how linguistic entities function as "signs" to gives us a clue, Berkeley will claim, to understand how scientific statements can have general import while dealing only with particulars. That he himself viewed the theory of signs as of extreme importance in explicit in the *Alciphron* (or the *Minute Philosopher*. (Dialogue VII-16.) Euphranor (Berkeley's spokesman) remarks:

I am inclined to think the doctrine of signs a point of great importance, and general extent, which if duly considered, would caste no small light upon things, and afford a just and solution of many difficulties.

---

[7] The classic and historically crucial exposition of this position, is the "Fifth Meditation" of Descartes; "On the Essence of Material Things."

[8] *Principles* 124 makes this plain.

[9] The problem of the "infinite divisibility" of finite extensive segments, as we will argue, is an expression of an extremely important problem in the Philosophy of Physics; what is the justification for considering "physical space" as "dense" or "continuous" particularly when such attribution conflicts with the alleged structure of perceptual space.

Within the content of Berkeley's theory of meaning, and his theory of signs, we will have occasion to discuss his theory of abstraction; particularly what he views to be the result of illegitimate abstraction. We will conclude that his theory of signification many not have had the architectonic virtue he wished it to have; that problems with the meaning of general classificatory terms like "man" or "triangle" are distinctly different from problems concerning the meaning of such terms as "light ray," "material substance," "space," "force," "velocity at an instant" and "infinitesimal." Berkeley tends to lump the misuse (as he considers it) of such terms together as instances of a mistaken belief in the existence of "abstract ideas."

Although the architectonic virtue may be lacking; Berkeley's discussion of specific problems is always instructive. He wrestled with the most serious problems in the foundations of mathematics and physics: how can we understand geometry, for example as both an empirical and an apodictic science; how explain the success of the calculus, given the conceptual incoherence in the concept of the "infinitesimal" or "instantaneous rate of change"; how is "gravitational force" to be understood; what is the nature of a "mechanical" explanation? Both with respect to methodological and substantive issues, Berkeley made a profoundly serious and critical attempt to come to grips with the Newtonian world view.

BERKELEY'S THEORY OF SIGNIFICATION

Two components of what has come to be called the "picture theory of meaning" [1] are that the non-syncategorematic terms of a language are meaningful if and only if they have sensible referents ("ideas" in the Berkeleyian sense) and those sensible referents are the "meanings" of the terms. A superficial reading of Berkeley would suggest that he holds a view somewhat like the above, but such as interpretation of his views would be completely misleading. There are, it is true, entries in the early philosophical notebooks which directly suggest such an interpretation, but the later works such as the *Principles of Human Knowledge, The Three Dialogues Between Hylas and Philonous* and the *Alciphron,* make it abundantly clear that this is not Berkeley's considered view.[2] It is therefore surprising that modern commentators still attribute to him that particular theory of meaning.[3] The error is a serious one, for it is precisely the Lockean or "picture theory" that Berkeley finds at the root of misunderstandings concerning the function of general terms in language, and at the root of misunderstandings concerning the nature of scientific knowledge. Berkeley has Euphranor tell us, in the *Alciphron,* that it is that conception of language which requires

that every substantive name marks out and exhibits to the mind one distinct idea separate from all others, (AL-VII-5).

[1] The historical reference here is to Locke, *Essay Concerning Human Understanding,* Book III "Of Words." To the extent that we present Berkeley's critique of what he understands to be Locke's position, no claim is made that he in fact has correctly understood that position.

[2] As we will point out in the text, it is not always clear whether Berkeley means by "inconceivable" with reference to a concept, "unintelligible" or "unimaginable." We will argue that often he means the first, and that the second of the term is logically derivative.

[3] For example, Popper attributes the following position to Berkeley: "To have a meaning, a word must stand for an 'idea'; that is to say, for a perception, or the memory of a perception; in Hume's terminology, for an impression or its reflection in our memory." Popper *op. cit.,* p. 444. Under these strict conditions the word "God" would be meaningless.

which is the source of the erroneous belief that men have the power to frame abstract general ideas.

Berkeley has Alciphron express the position he associates with Locke that linguistic communication requires not only that each non-syncategorematic term have corresponding to it a sensory content or "idea" for the speaker but that communication requires that the same train of ideas be elicited in the hearer. (or reader.) In the *Principles* he contends that it is this "received opinion" that is in error, the opinion that

language has no other end but the communicating ideas, and that every significant name stands for an idea. This being so, and it being withal certain that names which yet are not thought altogether insignificant do not always mark out particular conceivable ideas, it is straightway concluded that they stand for abstract notions. (*Principles*-sec. 19-Introduction) [4]

This view of language, that each "noun substantive" requires a unique sensible referent in order to be meaningful, would entail an ability to frame "abstract general ideas," otherwise such terms as "red," "man," etc., would be meaningless since they do not name or refer to one unique sensible particular (over and above the particular items in the extension of the class).

The two pillars, then, of the Lockean theory (for Berkeley) are that grammatical subjects must have sensible referents in order to be meaningful, and that "understanding" consists in an immediate translation of the auditory or written sign into its sensible referent. Berkeley easily confutes the latter contention. We often operate meaningfully with linguistic signs without conjuring up in imagaination sensible referents for those signs. Words, Euphranor points out in the *Alciphron*, can be used like counters in a game: although each counter stands for a determinate quantity we can use them without referring each time to the quantity to which they refer. The view was expressed vigorously, as well, in the *Principles*:

... names being used for the most part as letters are in algebra, in which, though a particular quantity be marked by each letter, yet to proceed right it is not requisite that in every step each letter suggest to your thoughts that particular quantity it was appointed to stand for. (*Principles*, Introduction – 19)

---

[4] It is important to remember that Berkeley is speaking here of communication; that is the conditions for a term to be *meaningful* to a subject. He is denying that some ideational referent for the term must be present in the communicator and the communicatee. We understand Berkeley to mean this denial to apply to both singular and general terms, although as a principle it is often applied in the discussion of the significance of general terms.

This syntactical emphasis, which views understanding in part as operating in accordance with rules for the combination of signs, although quite important for Berkeley, leaves open the possibility that a referential theory of meaning is still correct; meaningful non-syncategorematic terms are those that refer or "name" one and only one entity; moreover, the "entities" named are sensible entities. With respect to linguistic signs, Berkeley rejects this view of significance, which would reduce meaning to referring in the sense that proper names refer to or name a distinct object.[5] Limiting our discussion now to the role of general classificatory terms like "man" or "triangle," Berkeley's view is that such terms have "divided reference"; [6] that is, although they can refer to any member of a given class, they do not name a particular entity over and above those members. Berkeley would not deny that recognition of similarities among individuals is a precondition for classification. He would deny that such similarities constitute the reference for the general term. Although objects may be similar, similarities are not objects, particularly if we interpret as Berkeley does, "object" as determinate sensory content.

The view that "noun substantives" refer to particular sensory contents, would be false for Berkeley, even if the discussion were limited to singular terms. The term "God" for example, refers or names a particular spiritual entity, but does not refer to an "idea."

An agent therefore, an active mind or spirit, cannot be an idea or like an idea. Whence it should seem to follow that those words which denote an active principle, soul, or spirit do not in a strict and proper sense, stand for ideas. And yet they are not insignificant neither; since I understand what is signified by the term "I," or "myself," or know what it means although it be no idea, but that which thinks, and wills, and apprehends ideas, and operates about them. (*Alciphron* VII-5)

Moreover, some general terms like "force" are significant in physics, Berkeley claims, although no instance of the class of "forces" is an "idea" (the same would hold true of the general term "spirit").

Berkeley's early and continuing investigations in arithmetic and algebra give abundant evidence of his interest in a theory of signs. In the Alciphron, for example, he has Euphranor say:

[5] This is no more to say then that general terms are not proper names or individual descriptions. What constitutes an individual is not discussed by Berkeley and this is a serious lack in his work. Unique spatio-temporal locus (even "apparent" spatio-temporal locus which might be compatible with the doctrine of the ideality of space) would not be sufficient, for in Berkeley's ontology there are individual spirits as well as "things."

[6] The expression "divided reference," referring to the function of general terms is used by W. O. Quine, in *Word and Object,* (New York: John Wiley and Sons; 1960) 90-95.

As arithmetic and algebra are sciences of great clearness, and certainty, and extent, which are immediately conversant about signs, upon the skillfull use and management whereof they entirely depend, so a little attention to them may probably help us to judge of the progress of the mind in other sciences, which though differing in nature, design and object, may yet agree in the general methods of proof and inquiry. (*Alciphron* VII-12)

We will soon comment in detail on Berkeley's theory of "signs"; making some remarks now on how his interest in "signs" indicates something about his general theory of linguistic meaning.

We deal with the world, he believes, not always directly, but mediately, through the use of signs, or more generally language; and such an ability suggests for Berkeley a quite thoroughgoing critique of what he at least understood to be the Lockean theory of linguistic meaning: that to understand a term required that we have some percept (or image) of what it refers to. On the contrary, Berkeley argues, the human mind is not designed for the "bare intuition of ideas," but for "action and operation" about ideas; such action leading to human well-being or happiness. The measure of knowledge is practical mastery of particulars, not the intuition of essences. (*Alciphron* VII-11) Although Berkeley never denies (even in arithmetic) the semantic component in a system of signs – that they refer in their genesis and ultimate reference to non-linguistic entities – he insists that a necessary condition for understanding does not require attention to these referents, but merely attention to the rules for the combination of "signs" (syntax). No psychological awareness (as percept or image) of the referent for a term contemporaneous with language use is required for there to be understanding.

If I mistake not, all sciences, so far as they are universal and demonstrable by human reason, will be found conversant about signs as their immediate object, though these in the application are referred to things. (*Alciphron* VII)

Broadly speaking, progress in science is in gaining foresight, and this requires recognition of the "regularities" ("laws") between kinds of phenomena. Focusing on the referents for linguistically formulated "law statements" emphasizes the descriptive aspect of such statements; they allegedly say something about the world. Focusing on a set of law statements as *rules* for the combinations of signs focuses on what we may call their prescriptive aspect. The algorithms of arithmetic, for example, can be considered purely syntactically as prescribing how certain signs should be combined.

Viewing law like statements prescriptively, as rules for the production of signs, emphasizes the instrumental character of "laws," particularly the mathematically formulated laws of mechanics. When given the "law" for free fall near the surface of the earth, $S = 16t^2$, and a particular value for "t," I can determine or produce a new "sign," a value for "S." Although useless if not ultimately referred to "things," the operation can be performed without any reference to them. "The human mind," Berkeley says:

is designed for action and operations about ideas ... (and for this) certain general rules are required, the formulation of which requires an "apposite and skillful management of signs." (*Alciphron* 7-11)

Berkeley's central point is that simultaneous intuition of their referents in not required for the "skillful management of signs," merely correct application of the algorithms for their combination.

Rejecting a view of understanding or communication which requires a contemporaneous intuition (as percept or image) of the referent for a sign as a necessary condition for its significant use allows Berkeley to handle more easily the question of the significance of general terms. Since no term to be understood requires a simultaneous consciousness of its referent (as an image) there is less pressure for requiring that some mental content be present as an object of consciousness, when general classificatory terms are used, for example, in the formulation of general laws. The significance of a general term can be said then to be exhausted by its function of referring to any particular that has the property (p).

We must distinguish, however, a theory of understanding from a theory of meaning. In denying that some mental content is required to communicate either in singular or general terms, Berkeley is not, of course, denying that extra-linguistic reference of some sort is possessed by meaningful singular or general terms. General terms for Berkeley have no *unique* ideational reference [7] and it is possible that what enables him to take this view has more to do with his views on the nature of understanding language, than with some Theory of the conditions which make linguistic expressions meaningful. Particularly important, in this regard, is his rejection of the view that "understanding" is an act of attaching mental content to a linguistic "sign."

Berkeley does deny that a necessary condition for a non-syncategorematic term to be meaningful is that it have some reference to a datum of sense. It might be instructive then to ask whether we can discover

[7] Again this means there is no single object they refer to. It is however, too restrictive to limit the extension of "object" to sense objects. "Mind" would be a general term, but even in its divided reference it does not refer to sensible particulars.

criteria in his writings for situations when linguistic expressions are meaningless; a task perhaps simpler than the approach of seeking a precise criteria of significance. Berkeley will argue, for example, that the expression "material substance" is meaningless, but not the term "spirit"; that the numerical sign "2" is meaningful, but not the expression "absolute space," or "velocity at an instant"; the theological term "grace" is meaningful but not the term "force" as used by those who speak of an immanent and animating principle in matter. Are there criteria that would categorize such diverse expressions as "material substance," "absolute space" and "velocity at an instant" as all being devoid of meaning?

Berkeley offers essentially two criteria for the meaninglessness of a linguistic expression; (1) the definition is self-contradictory; that is that any object in the extension of the term would have logically incompatible properties; and (2) the term is vacuous; that is there is no criterion for identifying anything in its extension. Reference to (1) is found throughout Berkeley's writings. The point is made by Philonous in the Three Dialogues:

I say in the first place that I do not deny the existence of material substance merely because I have no notion of it, but because the notion of it is inconsistent ... Many things for ought I know, may exist whereof neither I nor any other man has or can have any idea or notion whatsoever. But then those things must be possible, that is nothing inconsistent must be included in their definition. (*Dialogues* p. 177)

In the *Principles,* again referring to the concept of "material substance" there is the following comment:

Strictly speaking, to believe that which involves a contradiction or had no meaning in it is impossible; and whether the foregoing expressions are not of that sort, I refer to the impartial examination of the reader. (*Principles* 54)

The point is reiterated in one of Berkeley's later works, where the reference is to the expression "instantaneous velocity."

I further desire to know, whether the reader can frame a distinct idea of anything that includes a contradiction. (*Defense of Free Thinking in Mathematics* – 46)

The last passage admittedly is ambiguous, possible suggesting that the important criterion for absence of meaning in an expression is that no "idea" can be "framed" of what the expression purportedly refers to. In the class of such "unconceivable" ideas would be those which would have logically incompatible properties. Our own judgment, particularly in view of the long discussion of meaning in the Alciphron, is that it is self-contradictoriness (from which the inability to "frame" an idea follows as a consequence) which is the fundamental criterion. And although Berkeley is not as precise as one might like about whether "inconceivable" means "unimaginable" or "logically incoherent"; the latter sense, we belive, is the primary one.

There are expressions, however, which, while not referring to entities with logically incompatible properties, are devoid of significance for Berkeley, because they are vacuous; there is no way, he contends, for ascertaining whether anything belongs in the extension of the expression. Berkeley has suggested that the expression "material substance," even if internally coherent is vacuous. He will argue that the expression "absolute space" and at least one usage of the term "force" are vacuous expressions and therefore meaningless. We will investigate these claims in later portions of the essay.

D. M. Armstrong has recently claimed [8] that Berkeley himself does not adequately distinguish the "sense" or meaning from the "referent" of a linguistic expression. The context is the criticism of "abstract ideas" and the reference is to the Introduction to the *Principles* (sec. 18). Berkeley comments here that:

whereas in truth there is no such thing as one precise and definite signification annexed to any general name, they all signifying indifferently a great number of particular ideas.

Armstrong suggests that by the "signification" of a term, Berkeley understands its meaning, and then Armstrong perhaps rightly accuses him of contending that general terms have no univocal sense. Moreover Armstrong locates the absurdity in what he calls the "lurking idea" of Berkeley's that "all words are names." Thus if "horse" names indifferently any one of a number of individuals in a class, then "horse" has a number of meanings. In our view Armstrong misreads the passage. "Signification" more likely means "reference" and not "meaning." In the passage referred to by Armstrong Berkeley distinguishes between the "definition" of a term like "triangle" ("a plane surface comprehended by three right lines") and the "signification" of the term.

It is one thing for to keep a name constantly to the same definition, and another to make it stand everywhere for the same idea; the one is necessary, the other useless and impractical. (*Principles*-Introduction-sec. 18)

---

[8] David M. Armstrong ed. *Berkeley's Philosophic Writings*, (Longon: Collier Books; 1965) p. 29.

Berkeley's point is in fact precisely the one Armstrong believes he should have made; that a term may refer separately to a number of individuals in a class and yet retain a univocal sense. Although Berkeley uses the phrase "general name" when speaking of general terms, there is no evidence that he views them as analogous to proper names, where it might more plausibly be argued that the sense or meaning of the term is exhausted by its reference. In fact, Berkeley's distinction between the definition and the signification of a term is in more modern terminology a distinction between sense and reference. Although it is true that necessary condition for the meaningfulness of a general term is that we can know in principle to what objects it refers, such objects do not, for Berkeley, constitute the meaning of the term.

Although we have suggested that internal consistency and non-vacuousness may be necessary conditions for the meaningfulness of linguistic expressions, there are passages in the *Alciphron* and the *Principles* which suggest a much broader behavioral criterion for meaning. For example, in the *Alciphron,* Euphranor concludes his criticism of the "Lockean" theory of meaning by saying:

But although terms are signs, yet having granted that those signs may be significant, though they should not suggest ideas represented by them, provided they serve to regulate and influence our wills, passions, or conduct, you have consequently granted that the mind of man may assent to propositions containing such terms, when it is so directed or affected by them, not withstanding it should not perceive distinct ideas marked by the terms. (*Alciphron Dialogue* 7 Sec. 8) [9]

Berkeley is speaking here of the theological term "grace" and claiming that the term, like the term "force" in mechanics, is meaningful, though neither refers to a particular idea or object of sense. Yet, whereas Berkeley will later argue that the term "force" in mechanics is really elliptical for "mathematical hypothesis," "grace" and other theological expressions are meaningful since they elicit a consistent pattern of behavior, or, in his own words, they serve to "regulate and influence our wills, passions, and conduct."

This broad criterion of meaning should not, however, be taken as a

---

[9] Also *Principles* 20:
"Besides the communicating of ideas marked by words in not the chief and only end of language, as is commonly supposed. There are other ends, as the raising of some passion, the exciting to or deterring from an action, the putting the mind in some particular disposition — to which the former is in many cases barely subservient, and sometimes entirely ommitted, when these can be obtained without it, as I think does not unfrequently happen in the familiar use of language."

sufficient condition for the meaningfulness of linguistic expressions. If a patterned behavioral response to an expression was such a condition, the term "material substance" might, along with "grace" and the "triune God," turn out to be meaningful. The fact that a term referred to an internally inconsistent concept would not preclude its eliciting some consistent behavioral response. Rather we should take Berkeley as meaning that the significance of linguistic expressions can be measured not only by their referential function, but by their power to regulate the affective component of human life. On another level, the significant analogy between "force" and "grace" is that such terms refer to that which cannot be ascertained (referred to) independently of its effects. We will have occasion to discuss this issue in relation to "force," raising the question whether "force" *means* nothing more than certain "effects"; or whether such effects are evidence for the presence of "forces" where the latter is conceptually (if not referentially) distinct from force-effects.

It is difficult to find in Berkeley's writings a precise formulation of the relation of signification (X is the sign of Y). Formulations like "X stands for Y"; "X brings Y to mind" (through the mechanism of association), "X directs our attention to Y," capture something of what is meant by "X signifies Y." In all cases however we can speak of a sign vehicle (the "sign" proper) or that which consciousness is immediately attending to, and the designatum or that which is signified. The broad scope Berkeley attributes to the sign relation is expressed in numerous places in his work. *In the Theory of Visual Language,* for example, he says:

Ideas which are observed to be connected with other ideas come to be considered as signs, by means whereof things not actually preceived by sense are signified or suggested to the imagination, whose objects they are, and which alone perceives them. And as sounds suggest other things; so characters suggest those sounds; and in general all signs suggest the things signified, there being no idea which may not offer to the mind another idea which hath been frequently joined to it. In certain cases a sign may suggest its correlate as an image, in others as an effect, in others as a cause. But where there is no such relation of similitude or causality, nor any necessary connexion whatsoever, two things, by their mere co-existence, or two ideas, merely by their being perceived together, may suggest or signify one the other, their connexion being all the while arbitrary; for the connexion only, as such that causeth this effect. (*Theory of Visual Language* Sec. 30)

This catholicity concerning what can be included in the sign relation is echoed in the *Alciphron,* where more stress is placed on the function of non-natural signs: (linguistic expressions, diagrams, etc.)

We substitute things imaginable for things intelligible, sensible things for imaginable, smaller things for those that are too great to comprehend easily . . . present things for absent, permanent things for perishing, and visible for invisible. Hence the use of models and diagrams. Hence lines are substituted for time, velocity and other things of very different natures. Hence we speak of spirits in a figurative style, expressing the operations of the mind by allusions and terms borrowed from sensible things, such as apprehend, conceive, reflect, discourse and such like . . . (*Alciphron* Dialogue 7 Sec. 13)

With respect to non-natural signs, algebra is the paradigm case for Berkeley, for it most clearly exemplifies the stipulative character of all linguistic expressions; the lack of any "necessary connection between sign and designatum. Neither "similitude" nor "causality" are involved, merely the decision to let one "idea" stand for or represent something else. With regard to linguistic expressions, there is an ultimate decisional component as to what they refer, an element perhaps most apparent in algebra.

Perhaps of more importance, algebra exemplified for Berkeley, the significance of the syntactical or structural component of language; that is the "rules" which tell us how elements of the language can be properly combined. And the question of proper or improper combination can often be considered quite apart from questions concerning the "conventions of reference"; that is questions concerning what the elements of the language refer to. Algebra exhibits with a kind of purity our lack of concern with what if anything the individual variable or constant signs refer to. And if one treats algebra only as a formal scheme, then understanding it reduces to understanding the "syntax," the rules for the combination of elements.

It would be mistaken, however, to make too much of Berkeley's "formalist" tendencies. Even in his Philosophy of Arithmetic, as we will later argue, he never denies that the elements of arithmetic have some extralinguistic reference. What algebra exhibits (in the same way that using "tokens" for money in card games exhibits this) is that we can correctly use (and in that sense "understand") language without attending to the referents for individual elements.[10]

Although Berkeley's discussion of "signs" mentions both natural and artificial signs, he often chooses to discuss the structure of signification

[10] Berkeley will allow a purely "formal" sense to algebra which he does not, however, allow for arithmetic. Berkeley never denies that arithmetic deals with actual collections, although its algorithms can be performed without attention to those collections.

in terms of language; his suggestion that natural phenomena constitute a "language of nature," and that science articulates the "grammar of nature" occupies an important place in his work, though if pursued beyond mere metaphor, leads, as we will argue, to serious confusions.[11] Among modern commentators, Colin Turbayne has most consistently stressed the importance of the language "metaphor" in Berkeley's philosophy of nature.[12] He writes:

Descrates' "giant clock work of nature" and Berkeley's "universal language of nature" identify two opposing world views, one of which is well known, the other not, and correspondingly two opposing metaphors or models, one of which has been used with enormous success in the sciences, and the other barely tried.[13]

Unfortunately, Turbayne does not suggest how, in fact, Berkeley's "model" or "metaphor" should be applied in the solution of scientific questions. One suggestion of course, is that if one were operating with Berkeley's model one would not seek hidden causes "behind" the phenomena and be content with describing the regularities occurring within the observable world. If, however, by "hidden causes" is meant "occult" causes or imminent and vital springs of motion, Berkeley is not to be distinguished from Newton (or Descartes) in rejecting such causes. Another suggestion is that the two models represent the opposition between corpuscular theories which posit theoretical and unobservable entities, and macroscopic theories (like gravitational theory or classical thermodynamics) which remain on the plane of what is directly observable. Yet, as we will see, the relation between Berkeley's thought and "corpuscularism" is not as simple as may appear at first glance; and that an instrumentalist interpretation of "microscopic" theories might be compatible with his views.

[11] In the first edition to the *Principles* (108) Berkeley says:
"That the steady consistent methods of nature may not unfitly be styled the 'language' of its 'Author,' whereby he discovers his attributes to our view and directs us how to act for the convenience and felicity of life. And to me, those men who frame general rules from the phenomena, and afterwards derive the phenomena from those rules seem to be grammarians, and their art the grammar of nature."
Later in the same edition (110) and referring to Newton's *Principia*, he remarks: "The best grammar of the kind we are speaking of will be easily acknowledged to be a treatise of mechanics, demonstrated and applied to nature by a philosopher of a neighboring nation whom all the world admire." See Turbayne, *Principles Dialogues and Correspondence, op. cit.,* p. 75, 76. These locutions drop out in the second edition, although the concept of the phenomena of nature as a language is discussed in Berkeley's last major work, *The Siris.* It is a central theme of the works on vision, that visual data constitute a "language" whose referent is the data of touch.
[12] See also G. W. Ardley *op. cit.,* Sec. IV "The Analogy of the Grammar of Nature."
[13] Turbayne ed. *Works on Vision op. cit.,* Introduction xxvii-xxxvii.

Stronger support for Turbayne is obtained if we consider the Cartesian model as suggesting that there exist necessary connections between the hidden mechanism and the observable phenomena, whereas the language model suggests that the relations between phenomena expressed as laws of nature are contingent, as in the relation between a sign and its designatum. Locke himself never quite eschewed as an ideal (if not a real posibility) for scientific knowledge, knowledge of the *necessary connections* between the corpuscular structure of matter (its "real essence" and its observable properties.) [14] However, since within his theory of perception "observable properties" became "ideas" or contents of consciousness, the scientific ideal becomes explanation of how the interaction of material objects and the materially construed nervous systems of sentient beings can necessarily produce certain mental states. This alleged causal dilemma, as we will see later, tends to obscure the function of "theoretical" entities in atomic theories, and as previously mentioned the theoretical use of such entities may not be incompatible with Berkeley's views.

Within the context of the issue of "necessary connection," between phenomena, Turbayne is undoubtedly correct in his view that the crucial aspect of the "language model" is the concept of "ostensive definition" or what we have called the "conventions of reference." [15] Given even the vagaries in the concept of "necessary connection" [16] there is clearly

[14] For example: Locke, *Essay Concerning Human Understanding*, (Chicago, Henry Regnery Co., 1956) p. 102.
"Had we senses acute enough to discern the minute particles of bodies, and the real constitution on which their sensible qualities depend, I doubt not but they would produce quite different ideas in us, and that which is now the yellow color of gold would then disappear, and instead of it we should see an admirable texture of parts of a certain size and figure." A good discussion is found of the problems in Locke's view in Buchdahl *op. cit.*, Chapter 4.

[15] Turbayne, *Berkeley, Works on Vision, op. cit.*, Introduction p. xxx. Turbayne accepts rather uncritically Berkeley's views concerning the language "model." He quotes from the *Essay on Vision*, (140) "it is customary to call written words and the things they signify by the same name," p. xxxiv. It seems mistaken, however, to call our verbal locutions "names" of the written characters, as opposed to their pronunciations. The point is of some importance, since Berkeley uses the example as an analogy to our calling certain visual cues by the same name as their tactual designata.

[16] Two conceptions of "necessary connection" play a role in Berkeley's thought. The first, which we might call "analytic necessity" refers to the necessary connection between the axioms and a theorem in Euclidean Geometry. (See *Essay on Vision* 5). The meaning of "necessity" here is clearly "logical entailment." The second sense is "causal" or "productive" necessity; for example, the necessary connection between the divine volition and God's creation. If we assume God willed that "a" occur, it would be contradictory to assert "a" will not occur. There is of course a suppressed premiss; that it is impossible that what God wills not occur. Often for Berkeley, however, the suppressed premiss of divine omnipotence is not even operative, when Berkeley suggests that the necessity in the relation between volition and its effects

no such connection between a word considered as a sequence of marks or sounds and its referent, since it is ultimately a matter of choice what will be the designatum for a linguistic sign. The crux of the language metaphor, then, is to express the contingent nature of the relations between kinds of phenomena; relations that are ultimately formulated as "natural laws."

Berkeley, although he distinguishes between mere sequences of signs and sequences that can be said to constitute a language, considers the concept of the "language of nature" to be more than mere metaphor. In the *Alciphron,* for example, he writes:

That there are signs is certain, as also that language and all other signs agree in the general nature of signs, or so far forth as signs. But it is certain that all signs are not language: not even all significant sounds, such as the natural cries of animals, or the inarticulate sounds and interjections of men. It is the articulation, combination, variety, copiousness, extensive and general use and every application of signs (all of which are commonly found in vision) that constitute the true nature of language. Other senses may indeed furnish signs; and yet these signs have no more right than inarticulate sounds to be thought a language. (*Alciphron* Dialogue 4 Sec. 12) [17]

We find a similar view, more compactly expressed, in *Visual Language.*

A great number of arbitrary signs, various and apposite do constitute a language. If such arbitrary conventions be instituted by men, it is an artificial language; if by the author of nature it is a natural language. (*Visual Language* Sec. 40)

Let us for the moment accept uncritically the contention of the *Essay on Vision* that the immediate ("proper") objects of sight are merely light and colors, and that certain other apparent visual properties such as distance and magnitude are not "immediately seen" but are in reality tactual sensations (Berkeley includes kinaesthetic sensations) immediately brought to mind or suggested by the visual signs.[18] Yet the view that the visual cues to the tangible nature of objects constitute a language leads, if pushed beyond metaphor, to insuperable difficulties.

With respect to language utterances or expressions we can generally

(even for human wills) has something to do with the power of volition. Hume later clearly pointed out that we cannot construe any necessary relation between volition and its effects (for example, bodily movements) and there is certainly abundant evidence that the "will" to do "a" is not always followed by the occurrence of "a."

[17] *Essay on Vision* (147)
"Upon the whole, I think we may fairly conclude that the proper objects of vision constitute a universal language of the author of nature, whereby we are instructed how to regulate our actions in order to attain those things that are necessary to the preservation and well being of our bodies, as also to avoid whatever may be hurtful and destructive of them."

[18] We will later analyze this contention in more detail.

speak of two types of error; (1) misunderstanding the conventions of reference (labeling dogs "cats" for example), or (2) though understanding the conventions of reference, asserting falseley that something is the case. Comparable to incorrect usage and falsity of assertion we can meaningfully speak of correct usage and true assertion. It seems questionable whether it is meaningful, with respect to visual signs to speak either of correct understanding of the conventions of reference, or of true or false assertion. Natural signs, as a recent writer has suggested are perhaps fallible or infallible but not correct or incorrect.[19] Take, for example, an optical illusion where an expected tactual experience does not materialize. (One that has been suggested by the visual cue.) Here it is sensible to say, if we wish to stick to the metaphor, that the conventions of reference (associations between visual and tactual data) were more complicated than first thought, rather than being incorrectly used. Moreover, since we are not *in principle* able to know the "divine" conventions, we cannot *in principle* judge the truth or falsity of visual assertions. This is because any false prediction about expected tactual experience on the basis of visual cues would, again if we keep to the metaphor, be attributed to not understanding the conventions of reference. What God asserts to be the case would have to be the case; our failures are failures in understanding what he is asserting. That is, we can never in principle assume the fixity of the conventions of reference and raise the issue of truth.

The language analogy is even more difficult to adhere to if we speak of all phenomena (not merely visual) as a language of nature, and natural scientists as "grammarians." This is the position suggested in the *Siris*.

The phenomena of nature which strike on the senses and are understood by the mind, form not only a magnificent spectacle, but also a most coherent, entertaining, and instructive discourse; and to effect this, they are conducted, adjusted, and ranged by the greatest wisdom. This language or discourse is studied with different attention and interpreted with different degrees of skill. But so far as men have studied and remarked its rules, and can interpret right, so far they may be said to be knowing its nature.[20]

Here, it could be argued, there is no semantic component at all; there is no extra-linguistic reference. (Whereas with respect to the concept of "visual language," one might speak of the tactual properties of objects

---

[19] J. L. Austin, "Truth as Correspondence," in *Philosophical Papers*; (Oxford: The Clarendon Press, 1961) pp. 85-101. Reprinted in *Encounter, An Introduction to Philosophy*, ed. R. Corimer, E. Chinn, R. Lineback, W. Frankena; (Glenview, Ill.: Scott Foresman and Co.; 1970) 495-506.

[20] *Siris* Sec. 254.

as referents for the visual "signs.") Language, in principle, is constituted both by syntax and reference, but if the phenomenal world has been reduced to a congeries of "signs," the second component of language, or reference, is absent. It is not sufficient to suggest that the structure of the phenomenal world refers us to the divine purpose, for even if offering evidence for such purpose, it does not refer to it as a linguistic item refers to its referent.

Berkeley often appears to be focusing on the syntactic rather then the semantic component of the analogy. The suggestion that natural scientists are comparable to "grammarians" might be taken to mean that they articulate not the conventions of reference (semantics) but the syntax (grammar) expressed in the regular connections between types of phenomena; connections we call "natural laws." And just as with ordinary grammar or syntax we can distinguish "knowing how," from "knowing that"; (using as opposed to articulating the grammatical rules), we can distinguish practical from propositional knowledge with respect to natural laws. In the *Principles* (2nd ed.) Berkeley writes:

Those men who frame general rules from the phenomena and afterwards derive the phenomena from those rules seem to consider signs rather than causes. A man may well understand natural signs without knowing their analogy or being able to say by what rule a thing is so and so. (*Principles* 108)

It is important, however, to distinguish the "arbitrariness" (i.e., lack of necessary connection) of syntactical connection from the arbitrariness in the conventions of reference. The first, we might say, refers to the structural relations among linguistic items themselves, and it is this that might be metaphorically modeled in the conception of the "language of nature." Difficulties ensue when we seek to pass beyond the conception of the order among "signs," (i.e., regular connections among types of phenomena) to anything within nature that would model conventions of reference. Again, since in principle we cannot assume the fixity of such conventions and raise the issue of truth, there appears no point beyond metaphor of speaking of natural phenomena as a language.[21]

[21] The point can be made in the following way. If we wish to consider the data of vision, for example, as linguistic items, the "laws" which relate them as "signs" to referents (tactual data) are by Berkeley's own account synthetic. As are always in the process of learning what the referents for the visual linguistic items are. Any mistakes we make; optical illusions, etc.; refers to our lack of complete understanding of to what the visual items refer. Human errors in prediction therefore must always be attributed not to the falsity of a divine utterance, but to our lack of understanding the divine conventions of reference. This notion of a never completed learning of the divine conventions of reference is suggestive of the synthetic character of natural laws, but also suggests that there is little sense in speaking of natural phenomena as a divine language.

The metaphor does, however, focus on an important aspect of the structure of signification, whether of artificial or natural signs. The designatum is not necessarily or "immutably" connected to its "sign" although it may be invariably associated with it. The concept of "necessary connection," is, however, ambiguous in Berkeley's writings. On the one hand it connotes the connection between the theorems and the axioms in, for example, Euclidean Geometry, and on the other hand its sense is expressed by the connection between the divine volition and the divine effects. It is this second sense that is important here. Just as the reduction of phenomena to "ideas" expresses the lack of efficient causality in the former, the reinterpretation of causal laws as invariable associations between signs and their designata expresses a comparable lack of necessary connection. This is the meaning of the suggestion in *Principles* 108, that those "who frame general rules . . . seem to consider signs rather than causes."

The theory of signs, then, and the immaterialists metaphysics are two perspectives on the same theme: the lack of efficient causality within nature. And there are difficulties in Berkeley's conception quite comparable to the difficulties in Hume's later interpretation of the "causal" relation as an invariable association of spatio-temporally contiguous events. We will deal with some of these difficulties below.

We will now consider in some detail two areas where Berkeley's theory of signification plays an important role: (1) his analysis of the concept of "causality" and (2) his analysis of the significance of general terms.[22]

[22] There is, of course, another area where the theory of signification plays an important role; in Berkeley's theory of vision. Berkeley argues that our apparent visual apprehension of distance magnitude and figure is really an apprehension of certain tactual data which have through experience become embodied as it were in certain visual cues or signs. Although we will have occasion to discuss some specific aspects of the theory of vision, this particular thesis will not be discussed in detail. Five generally formulated criticisms of Berkeley's view, however, are given here:

1. The ambiguity in the concept of meaning. One of Berkeley's early critics, Samuel Bailey, interprets "meaning" to be "referent" and rightly points out that in the visual apprehension of distance no tactual referents (tactual sense data) are brought to mind. I can in principle pay attention to an image of the moon upon hearing "moon" but there are no tactual items (sense data) that can be brought to mind as distinct from certain visual cues. Berkeley's own analogy to language obscures the issue. When a word or sentence appears meaningfully to me, it is true I am not just paying attention to the perceptual item, linguistic or auditory. But it is not clear in what sense I am paying attention to "meanings" that are said to be embodied in these vehicles. Berkeley confuses the phenomenological sense of meaningfulness, with the concept of bearer or carrier of meaning.

2. Difficulties in the application of the psychology of association: Again Berkeley's early critics Bailey and Abbott point out in some detail that generally when we speak of a sign "a" bringing to mind "b" we are never under the impression that we actually perceive "b" with the same sense that perceives "a." In finding a musical score meaningful, I am never under the impression that I am seeing the music. Yet in

CAUSALITY

Apparently, Berkeley allows only one legitimate sense of "efficient causation," the production of "ideas" through the agency of will; either those "ideas" which in sum constitute the phenomenal world produced by divine volition, or the production of "images" through the agency of finite wills. This view not only has its roots in theological considerations, but in the radical break with Aristotelianism in sixteenth and seventeenth century natural science. The elimination of substantial forms or immanent principles of changes explanatory principles, that is, the giving up of the notion of an animating power *within* matter, raises questions

visual experience it is a distinct impression that the distance of objects is apprehended visually. Moreover if tactual experience can change the character of what I appear to "see," how explain optical illusions, where, for example, no matter how much tactual experience I have the stick in the water will appear bent.

3. Confusion between what we can call a genetic and a phenomenological account of visual experience. Although it may be true that tactual and kinaesthetic experiences are involved in the genesis of we appear to "see," we need not conclude that we are paying attention to tactual (or kinaesthetic) "meanings" for certain pure or "uninterpreted" visual data.

4. Difficulties in the concept of the "proper object of vision" itself. Berkeley's argument that "light and colors" at no distance from the eye constitute the immediate or "proper" object of vision suffers from confusion between genetic, phenomenological, and epistemological considerations. It is not clear whether he is speaking of what is in the visual field before any other type of sense experience (genesis) or that which at all times can be said to be properly in the visual field (as opposed to something not in the field but whose existence is entailed by a certain judgment.) (phenomenological – as when I say I "see" a man, which entails that the object referred to has a heart, an object not in the visual field.) The content of the phenomenologically "proper" visual field may not be the same as the content of the genetically "first" visual experience. There is a third epistemological sense of "proper" in which we might speak of incorrigible claims about what we "see." Historically the third was associated with sense data claims (e.g., "I see a red patch.") As opposed to physical object claims. ("I see an automobile.")

5. Failure to carefully distinguish metric judgments of distance and magnitude, from the alleged visual apprehension of distance and magnitude. Although there may be a relation between perceptual and metric judgments of distance, they are not the same. We will have occasion to discuss this issue in the text of the essay, pointing out, for example, that the fact that we have no innate ability to make metric judgments of distance, does not entail that "spatiality" or "outness" is not a "proper" object of sight in the genetic as well as the phenomenological sense of "proper."

It should be pointed out that passages in the *Essay on Vision,* are quite at variance with the theory of the significance of terms of the *Alciphron* and the *Principles.* For example, *Essay* 51: "No sooner do we hear the words of a familiar language pronounced in our ears, but the ideas corresponding there to present themselves to our minds; in the very same instant the sound and meaning enter the understanding." It is not clear in what sense we are to understand these "meanings." If "meaning" here has the sense of sensible "referent," then we have a view here explicitly rejected in the *Alciphron* and the *Principles.* Section 17 of the *Essay,* suggests this is, in fact, Berkeley's meaning. "Just as upon hearing a certain sound, the idea is immediately suggested to the understanding which custom had united with it." Yet as the *Alciphron* and *Principles* makes clear the sensible referent (qua image) does not need to be present for there to be understanding of the term.

concerning not only the cause of change, but the cause or ground of the permanence of objects; their continued existence in time. Descartes' conception of a continuous creation vividly expresses on the one hand the "passivity" of matter, and on the other, that the conservation of the world through time requires the continual agency of the divine will.[23]

In a reply to his American correspondent, Samuel Johnson, Berkeley explicitly makes use of the concept of a continuous creation.

Further, it seems to me that the power and wisdom of God are as worthily set forth by supposing him to act immediately as an omni-present, infinitely active spirit, as by supposing him to act by the mediation of subordinate causes, in preserving and governing the natural world. A clock may indeed go independent of its make or artificer, inasmuch as the gravitation of its pendulum proceeds from another cause, and that the artificer is not the adequate cause of the clock; so that the analogy would not be just to suppose a clock is in respect of its artist what the world is in respect of its creator. For ought I can see, it is no disparagement to the perfection of God to say that all things necessarily depend on Him as their conservator as well as their creator, and that all nature would shrink to nothing, if not upheld and preserved in being by the same force that first created it.[24]

The very existence of the natural order is intelligible only as a continuous effect of divine volition. Divine causality functions as a principle of intelligibility for the very existence of the natural order. It supplies a "necessary connection" lacking within the natural order itself. Particularly apposite here is a passage from *De Motu*.

It will be of great importance to consider what properly a principle is, and how that term is to be understood by philosophers. The true, efficient and conserving cause of all things by supreme right is called their fount and principle. But the principles of experimental philosophy are properly to be called foundations and springs, not of their existence, but of our knowl-

---

[23] Berkeley gets into some conceptual difficulty when, unlike the Occasionalists, he contends that human wills have causal efficacy. A passage in the early *Philosophic Commentaries* (548) reads: "We move our legs ourselves, tis we that will their movement. Herein I differ from Malbranche." However, if immaterialism is correct the "movements" of our bodies are sequences of "ideas" in some mind. Yet according to Berkeley, the objective order of the world, of which my bodily movements are a part, is the sequence of "ideas" impressed on numerous minds by God. The "public" character of objective knowledge is accounted for by the divine impression of "ideas" on a multiplicity of minds, as opposed to the essential subjectivity of self-produced images, dreams, hallucinations, etc. Yet Berkeley's rejection of Occasionalism would imply that finite minds had this power to impress not only upon its own mind, but on others, certain sequences of signs, which are part of the "objective" order of nature.

[24] In *De Motu*, Berkeley makes use of the principle of inertia to establish the same point. From a dynamic point of view, there is no distinction between rest and uniform motion; both represent mere "persistence," or "continuance in the same way of existing." (*De Motu* 27) Ultimately this "persistence" is explained in terms of the continual causal efficacy of the divine will.

edge of corporeal things, both knowledge by sense and knowledge by experience, foundations on which that knowledge rests and springs from which it flows. (*De Motu* 36)

Productive, or efficient causality, then is required to account for the existence (both the original creation and the continuance of objects) of phenomena, but is not an explanatory principle within physics itself. And, although we will note later that Berkeley's attribute of efficient causality to will (even finite wills) is grounded often in an alleged intuition of volitional "power," the attribution of causal efficacy to the *divine* will, functions as a principle of intelligibility, and what it makes intelligible is the origin and continued existence of phenomena.

How, then, should we understand the concept of causality applied not to the existence but to the order of nature? The world appears to us, Berkeley suggests, as if certain mechanisms

hidden as it were behind the senes (operate) in producing those appearances which are seen on the theatre of the world. (*Principles* 64)

The alleged dilemma, for Berkeley, is an illusory one. The relation between the "mechanism" and its "effects" is more properly construed as a signifying relation between two sets (types) of "ideas" the first labeled "cause" and the second (the designata) labeled "effect."

The "works" of a clock are perhaps rigorously correlated with the movements of the hands (*Principles* 62) but the former do not constitute an efficient cause of the latter. Although the distinction between "mechanism" and "effects" may be important in practice, it does not take us beyond the realm of "ideas"; the "works" of the clock qua "ideas" or objects of sense, are as passive as the movements of the hands. Berkeley then feels himself compelled to respond to the question as to the purpose of the elaborate mechanisms in nature and art if, in fact, there is no causally necessary connection between them and their effects.

To all which my answer is, first, that the connection of ideas does not imply the relation of cause and effect, but only of a mark or sign with the thing signified. The fire which I see is not the cause of the pain I suffer upon my approaching it, but the mark that forewarns me of it. In like manner, the noise that I hear is not the effect of this or that motion or collision of the ambient bodies, but the sign thereof. (*Principles* 64)

A number of difficulties are immediately apparent in the above view. Firstly, we would normally say the fire had caused the pain, regardless of whether the phenomenon "fire" functioned as a sign; secondly, as Berkeley himself admits, either "cause" or "effect" can function as the sign or the designatum in the relation considered. The charred wood is a "sign" that there *had been* a fire because we expect such an "effect" to be related to such a "cause." This is as much to say that the relation of signification is not simply a translation for what we term a "causal" relation. Berkeley might have accused us of being overly critical here, responding that the function of the uniformities of nature (natural laws) is to enable us to predict what will happen or retrodict what has already occurred. Therefore the demand that the relation sign-designatum be temporally isomorphic with the relation "cause and effect" is too strong. All we need to say is that the statement "A causes B" means given the occurrence of "A," I can assume (be forewarned) of "B," or given the occurrence of "B," I can assume that "A" had occurred. Yet, although it may be a consequence of establishing a causal relation between (A) and (B), that either (A) or (B) may function as "signs," this function cannot exhaust the meaning of the causal relation. Many spatio-temporally contiguous phenomena where one type of event is considered the "sign" of another type are not considered causal sequences. The darkening of the sky is a sign of rain (which may invariably follow it) but is not a "cause" of rain any more than the weather forecast.

We have, then, with respect to Berkeley, a problem comparable to the one we have later with Hume, of distinguishing from the broader set of spatio-temporally contiguous and invariably related (associated) types of event a narrower set termed "causal" relations.[25]

Aside from the alleged causal efficacy of volition, one traditional suggestion is that causal relations be attributed in situations when there is a "communication" of motion; situations expressed by statements such as "the wind knocked over the lamp," "the water buoyed up the log," [26] etc. Berkeley will argue, however, particularly in *De Motu*, that there is

[25] Bertrand Russell has pointed out that the condition of spatio-temporal contiguity between "cause" and "effect," cannot be maintained. If the events are not simultaneous, there is always some interval between them (some spatio-temporal interval) entailing that there is always some other "event" (thinking of "event" here as a portion of space-time) between the one we label cause and the one we label effect. Russell's argument is that because there is always some temporal interval (not an "infinitesimal") between "cause" and "effect," some other event may "interfere" with the occurrence of the "effect." This way of putting it seems faulty, however, since the notion of "interfering" appears to assume the point at issue; the notion of an event which insures (as cause) the occurrence of another event. See B. Russell, "On the Notion of Cause, with Applications to the Free-Will Problem; *Mysticism and Logic* (George Allen & Unwin, Ltd.) 180-205. Reprinted in H. Feigl and M. Brodbeck ed. *Readings in the Philosophy of Science,* (New York: Appleton-Century-Crofts, Inc.) 387-408.

[26] See the interesting discussion of transitive verbs of this type, and their function in explanation in: "Explanation Without Laws," Jerrold L. Aronson, *Journal of Philosophy* LXVI No. 17, Sept., 1969.

no identifiable item in experience ("idea") that corresponds to the phrase
"*communication* of motion"; moreover, the concept has no explanatory
value in mechanics. The laws of mechanics, he will suggest, including
the laws of impact, can be reformulated purely in kinematic terms, without any need to refer to "forces." "Communication" of motion, if used
at all, would be elliptical for a kinematic description where motion ceases
in one body and begins in another.

Berkeley, then, does not suggest the specific difference that would allow
us to distinguish "causal relations" from the more extensive set of uniform
relations in nature. If we eschew efficient causation, except where volition
is involved, we are still faced with the problem of distinguishing those
properties that characterize "causal" relations within the phenomenal
world (as a type of relation between kinds of phenomena).

We have suggested that the concept of the "passivity" of ideas expresses
within the Berkeleian idiom the view that there is no efficient causality
with respect to relations between kinds of phenomena. Such a view tends
toward the Humean conception that mundane causality *means* no more
than an invariable correlation between types of phenomena spatio-
temporally contiguous; although Hume would include here the associations between efforts of will and bodily movements. With respect to the
relations among material objects themselves, the view that such objects
are no more then "collections" of "ideas" (objects of sense), suggests
again that even "impact" phenomena do not demonstrate any "communication" of momentum.

Difficulties ensue, however, when we attempt to be precise about the
characteristics of "ideas" that entail their passivity. The connection between the concept of "idea," and the concept of "passivity" is not, for
Berkeley, analytic or definitional. At least he offers introspective psychological evidence for the connection as well as for the connection between
"will" and "causal efficacy." In *Principles* (25), for example, he says:

All our ideas, sensations, or the things which we perceive, by whatever names
they may be distinguished, are visibly inactive – there is nothing of power
or agency included in them. So that one idea or object of thought cannot
produce or make any alteration in another. To be satisfied of the truth of
this, there is nothing else requisite but a bare observation of our ideas. For
since they and every part of them exist only in the mind, it follows that
there is nothing in them but what is perceived; but whoever shall attend to
his ideas, whether of sense or reflection, will not perceive in them any power
or activity; there is therefore no such thing contained in them. A little
attention will discover to us that the very being of an idea implies passiveness and inertness in it, insomuch that it is impossible for an idea to do
anything or, strictly speaking, be the cause of anything . . ."

However, the contention that "ideas" have no causal efficacy appears to be dramatically at variance with Berkeley's theory of signification, the fundamental claim of which is that one idea can bring to mind or suggest another idea previously associated with it. Written or auditory signs, in themselves insignificant, direct us to the meanings of the terms of which they are composed; visual cues immediately bring to mind tactual associations (Theory of Vision); certain physiognomic changes (e.g. reddening of the face) suggest rage or embarrassment. There appears no way of accounting for this without supposing that "ideas" ("objects of sense and reflection") have causal efficacy. What Hume later calls "habit" is, after all, witness to this causal efficacy of individual objects of sense.

Berkeley might have countered that our objection does not alter the fact that one idea cannot produce another idea, and that this is the sense of efficient causality referred to. The passivity of "ideas" is contrasted with the causal efficacy of will; for it is only the latter that can produce ideas. Yet is seems somewhat gratuitous for this sense to exhaust the meaning of "efficient cause." Consider a memory recalled upon the occasion of a particular perception. Although the percept does not produce the memory (qua object of reflection) that is, it does not produce the content of the memory it is a causal agent with respect to the state of consciousness termed "having the memory." To claim otherwise would make unexplainable the fact that the mind, when presented with a particular datum, directs its attention (in memory or imagination) to some other datum. If a child could expect to be burned by the fire without any previous association, the fact that such an expectation *never* occurs without previous experience of fire would indicate a gratuitiousness in the divine creator.

Again, it could be objected that we have merely moved from the plane of physics to that of psychology. Certain kinds of mental states (expectations, images, memories, etc.) which are responses to "stimuli" (signs) are invariably associated with certain kinds and intensities of past experience, ("conditioning"), and that there is no necessary connection between these past experiences and the responses associated with them.

If we construe "necessary connection" in the Humean manner as expressing an analytic connection between the subject and predicate of a proposition, the objection is sound. The laws of associationist psychology, like the laws of physics, are synthetic. Yet this objection, in a fundamental sense, misses the point. In the first place there is no passage where Berkeley explicity identifies causal necessity with "necessary connection" in the Humean sense.[27] The necessity in the causal connection cannot be construed simply as logical entailment. If we think of what for Berkeley is the paradigmatic case of the causality of will in relation to bodily movements (or the production of images, memories, etc.), it would seem clear there is no logical entailment between an effort of will (conceived as a mental phenomenon) and a particular bodily movement. The causal agency of will is evidenced in the experience of effort attendant upon an exercise of will. Therefore, although causal connections are necessary connections in the sense that *if* (A) (for example and act of will) is the cause of (B) (a particular bodily movement) the conjunction of (A) and (not-B) is impossible in the same circumstances; no causal proposition is analytically true.[28]

We have the problem, then, of justifying our claim that for a given phenomenon to function as a "sign" Berkeley must attribute causal efficacy to it. He himself continually suggests that phenomena qua signs are causal agents. Take, for example, the following passage from *Visual Language*:

In treating of vision, it was my purpose to consider the effects and appearances, the objects perceived by my senses, the ideas of sight as connected with those of touch; to inquire how one idea comes to *suggest* another belonging to a different sense, how things visible *suggest* things tangible, how present things suggest things remote and future, whether by likeness, by necessary connection, by geometrical inference, or by arbitrary institution.[29]

Disregarding for the time being "geometrical inference" and "necessary connection," where the use of the term "suggest" may be elliptical, how else are we to understand the term in the context of the passage, except

---

[27] There is a sense (which we discuss) of "causal explanation" where logical entailment is involved; the sense in which an event to be explained (the occurrence of the event) is logically entailed by general empirical laws and certain given conditions. Berkeley will also call this "mechanical explanation," and distinguish it from metaphysical explanations in terms of the efficient causality of the divine volition.

[28] That is, it is not part of the meaning of (A) that it is the cause of (B). One can interpret Hume's contention; that no assertion of a causal relation is necessarily true in a purely epistemological way. This views Hume's contention as claiming that any assertion of a causal relation may be mistaken. This view is compatible with accepting the causal principle, that all events have causes, and that *if* (A) is the cause of (B) then it would be logically impossible to find a joint occurrence of (A) and (not-B). A good discussion somewhat along these lines is found in; Edward H. Madden, "Hume and the Fiery Furnace," *Philosophy of Science*, Vol. 38 No. 1, March, 1971.

[29] Similar passages which suggest that signs have a "power" to evoke or suggest their designata are found in *Alciphron*, Dialogue 7. What Berkeley understands to be the distinction between "necessary connection" and "geometrical inference," is not clear, since the latter is a species of the former.

to infer that certain phenomena qua *signs* have the power or causal agency to evoke in certain situations on image of that which they have been associated with. The conditions for having this "power" are the linkages established by past associations; the cloud formation becomes a "sign" of rain after we have experienced their temporal connection.

Berkeley, of course, would admit much of this. The linkages established by past associations are the conditions for a given datum of sense to become a sign; the association in experience of the heterogeneous data of sight and touch, for example, make possible the evocation of "tangible" ideas upon the stimulus of visual cues. More broadly, it is the mechanisms of association that makes possible for him, all non-demonstrative inference, from the implicit judgments of visual perception generated by the invariable association of visual and tactual sense data, to the consciously articulated generalizations of empirical science.[30] Yet, such an admission would seem devastating to his contention that "ideas" are passive; for the mechanism of association not only confers upon the mind a disposition to respond in specific ways to certain stimuli, but confers a certain property on objects of sense; that of being a stimulus – a property it gains only after certain experiences have taken place. The bell evokes salivation in the Pavlovian dog only after the animal has experiences the association of bell ringing and getting food. To consider a sense datum a *stimulus* would appear to assign causal efficacy to it.

Again it may be objected that labeling a sense datum a "stimulus" is an elliptical way of stating that it is (in certain situations) invariably followed by a certain response in an organism. However, from Berkeley's point of view this objection would undoubtedly commit one to "occasionalism" on the plane of ideas; that is, on the occasion of a certain "sign" God (or perhaps myself) elicits from me a certain response. Such a view, however, would render gratuitous the need for an experienced association between (A) and (B) as a necessary condition for (A) to become a sign of (B). We should expect Berkeley here to offer a criticism comparable to the one he offers of traditional "occasionalist" doctrine. This was to the effect that the existence of a material world was theoretically superfluous in the "occasionalist" account of the relations between body and mind. Instead of requiring that God "move" matter on the occasion of a particular mental act (e.g. an act of will), we can dispense with

[30] Berkeley considers the law of gravitation, for example, to be a simple inductive generalization, though one requiring the more acute observation of the "philosopher" who notices the similitude in quite diverse kinds of events, in this case the movement of bodies towards each other. We will argue later that this is an over-simplification; in what sense, for example, can we say the planets move towards the sun?

matter and consider all bodily movements as sequences of "ideas," hence mental phenomena themselves. The crux of the objection is that "matter" has no causal efficacy with respect to mental phenomena, (a point granted by the occasionalists themselves) hence God theoretically could produce ("cause") directly all our "ideas." The point is given general methodological significance in the *Second Dialogue Between Hylas and Philonous.*

And if it pass for a good argument against other hypotheses in the sciences that they suppose nature or the divine wisdom to make something in vain, or do that by tedious rondabout methods which might have been performed in a much more easy and compendious way, what shall we think of that hypothesis which supposes the whole world made in vain. (Dialogue 2, p. 155)

Similarly one could argue that if the experienced association between two kinds of phenomena is not required for one kind to be a sign of the other it would be a violation of the divine economy to have such a condition. Converseley, if such association is considered to be a necessary condition for one type of datum to be a sign of another type then we must conclude that ideas or objects of sense can have causal efficacy. Simply put, past association between (A) and (B) confers on (A) a property it lacked previous to the experienced association; that of being a "sign" for a given "interpretant." This property is the power to evoke or call to mind (in memory or imagination) that which had been associated with it.

Berkeley, as we have suggested, grounds the possibility of inductive or non-demonstrative inference in the linkages established in experience between distinct kinds of phenomena. These linkages may be of quite different sorts; pattern similarity that allows for geometrical classification (association by similitude), regular temporal association of completely heterogeneous sense data, (the data of sight and touch) and even recognition of functional relations articulated in the mathematical laws of motion. In a broad sense, even the time a body has fallen during free-fall might be considered a *sign* of the distance it has fallen. Although stretching somewhat the usage of the term "sign" – for the "time" fallen does not immediately evoke a judgment of distance – yet the ability to recognize a certain functional relation between time and distance makes possible a determination of one variable based on knowledge of the other.

The broad sense of "suggest" is of one item of experience signifying something beyond itself. Each temporal slice of consciousness is not a consciousness denovo, but reaches out in behavior, memory, images, the unconscious inferences of habit, and the conscious process of inductive generalization, towards that which has been associated with it. This concept of the function of experience in conferring upon some objects of sense the property of being "signs" is fundamentally unintelligible unless we reject the notion of objects of sense as passive and well bounded conscious contents, and rather attribute to them causal efficacy.

Although efficient causality, for Berkeley, is rejected in physics, it returns, perhaps unwittingly, in his psychology of association. Verbs like "suggest," "evoke," "bring to mind" and their cognates share with verbs like "push," "pull," "buoy up," the idea of a transference of power. The strong statement in the *Principles* to the contrary,[31] the doctrine of the passivity of ideas is not compatible with Berkeley's theory of signs.

The requirement that phenomena qua signs have causal efficacy is not, however, rooted in an intuition into the efficient causality of sense objects, understanding intuition here as something comparable to our reflective awareness of an effort of will. Rather such attribution of causal efficacy is rooted in what we might term a principle of intelligibility: to make sense of the fact that the past association between (A) and (B) is a necessary condition for (A) to be a sign of (B).[32] After all, the divine benevolence might be better represented by a world where no child would touch a hot stove, regardless of its past experiences. Past experience must for Berkeley create that condition in the organism (disposition) which makes possible a sense datum becoming a "stimulus" (which is conceptually to attribute causal efficacy to it).

We must remember that Berkeley accepts and elaborates in important ways the view that perception is implicit judgment. From his point of view, a description, for example, of the "objects" in our visual field, would normally not mention the "proper" or "immediate" objects of

---

[31] *Principles* 25: "It is impossible for an idea to do anything, or, strictly speaking, to be the cause of anything." Buchdahl (*op. cit.*, p. 302) interprets "strictly speaking," to mean that it is logically impossible for an "idea" to have causal efficacy. Berkeley's own views are not as clear, however. The fact that he offers introspective evidence for the "passivity" of ideas suggests that for him the issue is to some extent an empirical one. On the other hand it would seem to be a matter of logic that we could not observe in (as a property of) an "idea" something analogous to an "effort of will."

[32] From a "behaviorist" point of view, the question of why the need for past associations (linkages) could be ignored much as the question of how motion is "communicated" could be ignored with respect to the laws of mechanics. We merely have a complex correlation between a conditioning process (C), a stimulus (S), and a response (R). However, this would not, as it would in mechanics, be satisfactory for Berkeley. What would have to be made intelligible is the requirement for a conditioning process in order for a given datum to become a sign. Implicit in Berkeley's entire analysis is that such intelligibility is found by viewing the conditioning ("association") as that which confers upon the bare datum the property of being a 'sign.' The datum then has a quality (broadly speaking) it lacked before.

vision, (visual sense data, i.e., light and colors), but rather mention such things as physical objects, their distances, magnitudes and figures. Distance, however, to take one example, is not for him a proper object of vision, but a tactual datum that has been invariably associated with visual items and inextricably linked by the understanding. Disregarding at present the great difficulties with this view, it commits Berkeley to the belief that after continual association between visual and tactual sense contents, the "proper" objects of vision serve as cues, awakening in consciousness an awareness of the associated item. Yet this entire mode of analysis which posits linkages in the understanding, viual cues, etc., is filled with traditional causal analysis; what has been removed from mechanics appears to return in psychology.

Before closing this discussion of causality, we should note one usage of the concept in *De Motu*. To give the "cause" of a phenomenon here is held equivalent to "demonstrating" it or logically deducing it from general laws:

For the laws of nature having been found out, then it is the philosopher's task to show that each phenomenon is in constant conformity with those laws, that is, necessarily follows from those principles. In that consist the explanation and solution of phenomena and the assigning their *cause,* i.e., the reason why they take place. (*De Motu* Sec. 37)

This use of the term "cause," however, is not normal usage for Berkeley; generally he contrasts a "mechanical" explanation as described above, with a "causal" explanation, where the reference of the latter is to efficient causation. We will have occasion later to discuss this concept of mechanical explanation, which appears clearly to have its source in Newton's discussion of method in the *Principia*.

GENERAL TERMS

The formulation of natural laws requires classificatory terms, or what we can call "general terms." Berkeley's well known view is that a term gains generality not by referring to a "general abstract idea," but rather because it can be used to refer to any item in a given class.[33] We can understand here a class to be a collection of items which have at least one property in common. And although Berkeley grants that knowledge is of "universal notions," "universality," he claims, does not consist "in the

[33] There appears to be ample evidence as A. A. Luce points out (his introduction to *A Defense of Free Thinking in Mathematics* p. 105) that Berkeley maintained his views on abstract ideas. See sec. 45-58.

absolute positive nature or conception of anything." We might understand this to mean that no particular item of consciousness can have the character of universality. Universality, Berkeley goes on to say consists

in the relation (a particular idea) bears to the particulars signified or represented by it; by virtue whereof it is that things, names, or notions, being in their own nature particular, are rendered universal. (*Principles* – Introduction 15)

Berkeley does not draw precise distinctions between questions of the genesis and questions of the significance (meaning) of general terms. With respect to the former, he clearly appears to acknowledge that we have the ability to recognize similarities in objects, and so group them into classes. This is a necessary condition for the possibility of having a linguistic item which "stands for" or designates any member of a given class.[34] Below we will discuss one attempt of Berkeley's to show "how" words can become general, but as we will see the discussion centers on the function in discourse of general terms, not on their genesis.

With respect to the question of significance: Berkeley denies that a general term refers to anything other than the individual items in a class. There are no universals *in re* or *in mente* referred to by general terms. And given his interpretation of "idea" or "object of consciousness" as a

---

[34] Locke's definition of an "idea" as "an object of thinking" (Bk. 2 Ch. 1 sec. 1 *An Essay Concerning Human Understanding*) includes, however, universals such as "man," "elephant," "army." It would be difficult given this broad extension, to interpret "idea" as an object of sense. Locke does speak of forming general ideas by a process of abstraction. "Ideas," he says, "become general by separating them from the circumstances of time, place, and any other ideas that may determine them to this or that particular existence. By this way of abstraction they are made capable of representing more individuals than one; each of which, having in it a conformity to that abstract idea, is (as we call it) of that sort." (*Essay* BK 2 Ch. III sec. 6) It is difficult to understand what is meant here by a process of abstraction. If we think of it in some senses as removing all the properties from an individual object until we are left with its generic properties, Berkeley rightly considers such a notion as unintelligible; if we remove all the properties that distinguish one triangle from another, we are left, not with generic triangularity, but with nothing. Such a literal view of Locke's meaning is not, however, necessitated by the text. Locke may mean that although the world is only populated with individuals, we can focus on them qua elements of a class. This "paying attention" to the generic properties is not much different from Berkeley's own view of legitimate abstraction as "selective attention." It is difficult to believe that Locke felt that in abstraction some new sensible object was being constructed, a generic object.

With respect to the question of the genesis of general terms, we have this passage from the *Essay on Vision*: "When upon perception of an idea I range it under this or that sort, it is because it is perceived under the same manner, or because it has a likeness or conformity with, or affects me in the same way as the ideas of the sort I rank it under. In short, it must not be entirely new, but have something in it old and already perceived by me. It must, I say, have so much at least, in common with the ideas I have before known and named as to make me give it the same name with them."

sensory content having a particular temporal locus in the stream of consciousness, his denial is undoubtedly correct. A particular or individual, according to Berkeley, is individuated with respect to those properties about which it can be individuated, an individual triangle, for example, must be either acute, equi-angular or obtuse. Triangularity itself is not an object of consciousness; which amounts to saying that triangularity is not an individual object of sense.

Philosophic support or criticism of Berkeley's views on abstraction often is directed at the traditional problem of the existence of universals, and although important, is not the direction we will take. Rather it is instructive, we believe, to consider the examples Berkeley uses in the *Introduction to the Principles,* where he first discusses how particular geometrical diagrams gain generality and proposes to use this as a model for how words may become general.

As we have mentioned, Berkeley does not really discuss the genesis of general terms (or general "ideas") but their function. A geometrical diagram, in itself a particular perceptual object, has "generality" "by being made to stand for all other particular ideas of the same sort." It signifies "indifferently" any other particular triangle. Generality is then one type of signification, a type, however, within the genus of artificial as opposed to natural signs. General terms share with proper names the characteristic that their referential function is based on an original stipulation. It is this decisional element at the root of establishing a particular as a sign which, as we have mentioned, distinguishes artifical from natural signs. It is this element which is expressed in the phrase "being made to stand for."

More precisely we could say that a particular functions as a natural sign when it serves to suggest, or bring to mind some other particular with which it has been associated. A particular is an artifical sign when it is used as a token or stand in for other particulars (the role of proper names, definite descriptions, etc.) or for particulars of a certain class (the role of general terms). All artificial signs can become natural signs of their stipulated designata. The reverse is true as well; an ideographic language might select a linguistic particular on the basis of some recognizable similarity to what it refers to. It is, however, the stipulation that it will refer to something that makes it an item of language.

If we focus on this stipulative character of artificial signs, (whether of signular or general terms), Berkeley's comment in the *Principles* (Introduction-12) that "by observing how ideas become general we may the better judge how words are made so," is seen to be vacuous.

If we focus our attention on the function of generality, the particular diagram in a geometrical demonstration is as much a general term as "triangle." That it is of "the same sort" as the particulars it designates is not required for it to have general significance. Linguistic entities themselves, as Berkeley points out, are particular objects of sense; the distinction between "$\triangle$" and "triangle" is merely that the former resembles the particulars it individually refers to. Observing how a geometrical diagram gains generality would give us no insight into how an linguistic entity does so; what ever puzzles are involved in the second case would equally well apply to the first.

The function of the analogy, however, is not so much to explain how general "ideas" are possible, but to show that particulars can have general significance (as idea or term) without having to refer to an "abstract general idea."

And, as the particular line becomes general by being made a sign, so the name "line" which taken absolutely is particular, by being a sign, is made general. (*Principles* – Introduction Sec. 12)

Berkeley's contention, perhaps, is that it is obvious that the diagrams used in geometrical demonstrations are particular, yet they are used to arrive at general theorems; theorems that are true of every particular "of the same sort." Realizing this, we become aware that we can have knowledge (which is of the universal) without requiring the existence of "abstract general ideas" as the subject of predication.

Berkeley, however, does not sufficiently distinguish the question of the general significance of particular diagrams from their use in the demonstration of theorems. As the latter bears directly on his conception of geometry, we will give it detailed attention when we turn to his philosophy of mathematics. Some preliminary remarks, however, are in order. The "general significance" of a diagram does not require that it share any similarity with the items it refers to. Sharing similarities, however, ("being of the same sort"), is important for Berkeley with respect to the use made of diagrams in the demonstration, since he believes that demonstrations (and constructions) are about the particular figure in the diagram. But we will raise the question whether such figures can be validly said to be the subject of the demonstration.

There is one further point about the problem of abstract ideas. Berkeley suggests that many of our conceptual difficulties in the formulation of mathematical and scientific propositions stem from an erroneous belief that "abstract general ideas" (perhaps generic entities or simply universals would be somewhat equivalent) are the subjects of predication in such propositions. And, opposing this view, he continually insists that in scientific laws the grammatical subject is merely a shorthand for distributively referring to a specific class of particulars. His claim, for example in the *Principles,* is that the laws ("axioms") of motion are not predicated of some abstract entity called "motion," but rather of *any* particular motion. Likewise the claim that "all extension is divisible" is not about something called "extension" *simpliciter* (something which is "neither line surface or solid . . . or of any determinate color"), but about any particular extension.[35] (*Principles* – Introduction Sec. 11)

It must be mentioned, however, that Berkeley's criticism of the use of such concepts as material substance, force, and instantaneous velocity, is distinctly different from his criticism of what he understands to be Locke's theory of abstraction and the significance of general terms. And the philosophic issues involved in the debate between nominalism and conceptualism are not particularly germane to the former criticism. Berkeley's claim *is more* than that there is no generic entity named by the terms, "force," "material substance," or "instantaneous velocity." His claim is that there are, as well, no particulars named by such terms. There are no individual instances of material substance nor individual examples (in mechanics) of force. Such terms do not even have divided reference. They are either strictly meaningless, vacuous or elliptical for proper but quite different expressions.[36]

---

[35] The example is somewhat strange, since Berkeley will argue that finite extended segments are not infinitely divisible. However, we can take the example as dealing with the generality and not the truth of propositions.

[36] For example, in *De Motu* (23) we find the following passage: "But if anyone maintains that the term body covers in its meaning occult quality, virtue, form or essence, besides solid extension and its modes, we must leave him to his useless disputation with no ideas behind it, and to his abuse of names which express nothing distinctly. But the sounder philosophic method, it would seem, abstrains as far as possible from abstract and general notions. (If notions is the right term for things which cannot be understood.)" Berkeley is speaking of the use of the term "force" to designate an immanent principle of motility in matter. This "abuse" if it is one, is quite different from the abuse of thinking that the term "triangle" refers to a unique object of sense over and above particular triangles. G. J. Warnock too easily, in our judgment, accepts the view that misuse of terms like "material substance," "force," or "instantaneous velocity," is based on an illegitimate abstraction comparable to the error which posits "general ideas" as the referents for general terms. For example, Warnock contends:

"A man who knows what these scientists knew quite well already knows what force and what motion are; the refusal to be satisfied with this is due merely to the notion that there must be someone thing – an "essence," and "idea," or some such entity – that is really named by the words 'force' and 'motion.' And this is a special case of the general mistake of assuming that all words are really names." (G. J. Warnock, *Berkeley,* (Baltimore: Penguin Books; 1953) p. 204.

Warnock fails, however, to point out that problems with terms like "force" are distinctly different from problems with terms like "triangle." It may be mistaken to say that magnets and planets exercise a "force" but if this is mistaken, it is a distinctly different mistake from claiming that there exist general triangles.

CHAPTER II

THE THEORY OF VISION

We now turn to a more detailed consideration of Berkeley's theory of vision.[1]

Berkeley himself considered the conclusions of the *Essay* important evidence for "immaterialism," that all objects of sense (and hence all material objects) were mind dependent. In the *Principles* he remarks:

For that we should in truth see external space and bodies actually existing in it, some nearer, other farther off, seems to carry with it some opposition to what has been said of their existing nowhere without the mind. The consideration of this difficulty it was that gave birth to my Essay Towards a New Theory of Vision, which was published not long since. (*Principles* sec. 43)

A similar view, although more strongly phrased was expressed by one of Berkeley's nineteenth century critics, Thomas Abbott. "There is indeed," he says,

"only one dogmatic system consistent with the Berkeleian theory of vision, and that is the Berkeleian Idealism." [2]

Our own reading of the Essay, however, leads us to a quite different conclusion: that the legitimate claims of Berkeley's work on vision are not only compatible with a realist metaphysics, but offer no evidence at all for an idealist one. By "legitimate claim" we understand Berkeley's contention that the judgment of apparent distance depends in part on experienced associations between visual and non-visual sense data, and that such judgments are not made a priori through the mind's possession of an innate ("natural") geometry. And although Berkeley's metaphysical claims for his conclusions are not a major topic of this essay,

[1] Foornotes referring to Berkeley's works on vision, refer to the Turbayne edition; *op. cit.*
[2] Thomas K. Abbott, *Sight and Touch,* (London: Longmans Green; 1864) introduction.

we will point out how such claims do in fact function (illegitimately) to support important contentions of the *Essay*, particularly the contention that there exist sensible minima.

From Berkeley's works on vision, we have selected three topics for discussion. (1) His critique of geometrical opics; (2) The so-called "vulgar error"; that is, how important in the *Essay* is the assumption of a "real" space external to the observer; (3) The concept of sensible minima.

GEOMETRICAL OPTICS

Berkeley rejects the claim of Kepler and Descartes that we can judge distance and magnitude by means of a "natural" or innate geometry; that built into the mind as it were, are the Euclidean axioms that together with certain data about the distance between he eyes allow us to compute our distance from the object.[3] According to Kepelr we make implicit use of the "distance measuring triangle"; ("Triangulum distantiae mensorium") that is a triangle composed of lines from a point source to the two eyes (the "optic axes") and the line between the eyes. In addition, according to this theory we can necessarily judge comparative distance by implicitly or unconsciously noting the angle made by the optic axes with the point source; the greater the angle necessarily implies the nearer the object. Berkeley correctly points out that we are not conscious of such "lines and angles" (or one might add "light rays") and therefore cannot be said to making such inferences concerning the distance (and magnitude) of objects.

Since, therefore, those angles and lines are not themselves perceived by sight, it follows, from sec. 10, that the mind does not by them judge of the distance of objects. (*Essay* 13)

Section 10 makes the claim that "it is evident" that "no idea which is not itself perceived can be the means of perceiving any other idea." "We do not," as he says in *Visual Language*, "see by geometry." (V. L. 32)

Berkeley's claim, however, is marred by a fundamental ambiguity in his concept of distance, an ambiguity that occurs throughout the *Essay*. If we mean by "distance" metric judgments of distance, of the sort a surveyor might make, it would seem possible in principle for such judgments to be made by an innate or "natural" geometry. At least it appears to be an empirical question. Admittedly it would not be "inference" if

[3] A brief account of this concept of a "natural geometry," is found in Turbayne, *Berkeley's Works on Vision, op. cit.,* xxi-xxvi.

by inference we understand a conscious process of reasoning, but Berkeley himself does not seem to require that we be reflectively aware of all our inferences or judgments. If, in fact, men's judgments of metric distance before they learned geometry invariably matched the results gained through conscious geometrical inference, it might be plausible to hypothesize some physiological embodiment of the Euclidean axioms, which enabled us, given certain stimuli, to make accurate judgments of distance.

Berkeley, it should be mentioned, often treats the question as an empirical one. If the "natural geometry" theory were correct, he points out, certain consequences would follow, which, in fact, do not; those born blind, for example, upon regaining sight should immediately and correctly judge the distances of objects. Moreover (and what for Berkeley amounts to an experimentum crucis) there is the "Barrovian case" where due to a lense placed between the object and the eye the rays of light from a point on the object may converge as they reach the eye. The case appears to subvert an accepted principle of geometrical optics, for the image according to such principles should appear, as it does not, behind the subject. (*Essay* 30) [4] In fact, as Berkeley notes, as we withdraw from the object behind the lens, thus increasing the actual distance, the apparent distance appears to decrease.

However, if by *distance* is meant the phenomenal or apparent characteristic of the "outness" or externality of objects – a characteristic of the visual (or visual consciousness) itself – it becomes analytically true to say "we do not see by geometry." [5] The relative distances of objects appear to be a quality of the visual field itself. It is not normal to express this quality of distance or "outness" in metric terms, although, having had some experience with measurement, metric locutions ("the object appears to be *ten feet* from me") might well be used to express our awareness of distance.

This ambiguity in the concept of distance is not simply an accident. It is rooted in Berkeley's claim that the quality of "outness" is not an "immediate" or "proper object" of sight. We will deal critically with this

[4] Berkeley takes the "accepted principle" from Barrow; he quotes from Barrows's *Eighteen Lectures* (London, 1669), Lect. 18. Turbayne, *op. cit.*, p. 27 (footnote)

[5] This appears to be the claim of G. A. Johnston, *The Development of Berkeley's Philosophy*, (New York: Russell and Russell Inc; 1965, first published 1923) Johnston's point is that from a logical point of view Geometrical Optics cannot be a theory of vision since possessed of an innate geometry the blind as well as the sighted could equally well judge distance. Johnston's remarks indicate the necessity for distinguishing metric judgments of distance, from the phenomenal sinse of "outness" or "spatiality," claimed as a quality of the visual field. Often Berkeley, as we will argue, does not carefully enough make this distinction.

claim in the section titled "the vulgar error." A consequence of the claim, however, is that the term "distance" appears to have, at least with respect to the proper objects of sight, no phenomenal referent. Yet it must have *some* referent, in order for there to be possible the associations between *distance* and such phenomenal qualities as muscular strain and faintness of image, associations which, according to Berkeley, account for our judgments of distance. That referent cannot be the apparent distance (or "outness") of objects, which appears to be an intrinsic part of our visual field, for Berkeley will argue that this apprehension of "outness" in truth is a judgment based on experienced associations between the proper objects of sight ("light and colors") and something called "distance." We are left, as we will see, with two possible candidates as referents for the term "distance"; "distance" as a metric concept, that is, as the object of direct or indirect measurement in space from the observer to the object; or "distance" as some tactual or combination tactual-kinaeesthetic sense datum. And, if we reject the view that apparent distance as an intrinsic quality of the visual field, is an "idea," we will be left with the difficult if not insuperable problem of demonstrating that distance as the object of measurement (direct or by geometric inference) can be reduced to some tactual, or combined tactual-kinaesthetic sense datum.

However the above problem is resolved, Berkeley offers a distinctly different criticism of the thesis that we judge distance by means of a "natural geometry."

The truth of this assertion (that we *do not* judge distance my means of lines and angles) will yet be further evident to anyone that considers those lines and angles have no real existence in nature, being only a hypothesis framed by the mathematicians, and by them introduced into optics that they might treat of that science in a geometrical way. (*Essay* 14) [6]

The passage is unfortunately too condensed and somewhat vague, but it does raise some interesting questions. Berkeley appears to be introducing the notion of a "theoretical entity," in this case the lines and angles of geometry, and "entity" that having no real existence, allows us to treat an empirical subject matter (optics) in a mathematical way. Moreover Berkeley appears to be contending that the objects of geometry are "hypothetical" as opposed to real entities, a contention that seemingly contradicts later assertions in the *Essay* (and in the *Principles*) that sensible extension is the object of geometry.

We will concentrate here on the first issue raised in the passage, the

[6] See also the *Siris* (250). "Mechanical Philosophers and Geometricians take mathematical hypotheses as real beings."

role of geometry in geometrical optics. If, in fact, the "lines and angles" of geometry have no real existence, how can we explain usefullness not only in judging distances, but in explaining any "optical" phenomena such as the length of shadows, the phenomena of reflection and refraction etc? Although Berkeley gives no detailed answer to this question, we can suggest in outline what is would be. With respect to either monocular or binocular vision, the "lines and angles" would refer respectively to "light rays" and the angles they make with the line between the eyes (this "line" empirically represented by a straight edge) or with each other at a point on the object. A passage from the later *Visual Language* supports this view:

To explain how he mind or soul of man simply sees is one thing, and belongs to philosophy. To consider particles as moving in certain lines, rays of light as refracted, or reflected, or crossing, or including angles, is quite another thing, and appertains to geometry. (*Visual Language* 43)

For example, in the alleged innate monocular judgment of distance, the "distance measuring triangle" is the figure consisting of a line corresponding to the width of the pupil and two light rays whose vertex it at a point in the object. Knowledge of the relative "divergency of the rays" (given the constancy of pupil width) allows us necessarily to judge whether one object is further from us than another. (*Essay* 6)

It seems clear that the principle that bridges the formalism of geometry with its "non existent" lines and angles and optical phenomena is the hypothesis of the rectilinear propagation of light. The latter can be viewed as containing both a material and formal component. The material component is the identification of "light ray" with straight line; the formal component is the contextual or implicit definition of straight-line in terms of the Euclidean axiom.[7] We cannot, of course, attribute this type of distinction to Berkeley. Moreover although Berkeley suggests, in the passage previously quoted, (*Essay* 14) a concept of "pure" geometry; at least of a geometry whose terms refer to nothing in existence, this view appears to be at odds with his later assertions that sensible extension is the object of geometry.

However, Berkeley does appear to accept the view that light is propa-

---

[7] The distinction here is between what we might call the syntactic and the semantic component of a theoretical term. The identification of "light ray" with straight line could be called a "coordinating definition," a "bridge principle," or an "epistemic correlation," depending on which literature in modern philosophy of science is referred to. The situation here is somewhat more complicated, since "light ray" itself is not in any simple sense an observable entity. It too, has to be partially interpreted in terms of certain observable phenomena like shadows, refractive phenomena etc.

gated in straight lines, for clearly it is this hypothesis which makes geometry useful to optics. It is not, then, the "lines and angles" simpliciter which function as "theoretical" entities, but lines and angles materially interpreted as light rays and their intersection. By "theoretical" we understand merely that the term is in part defined by axioms in a theory. (In this case, geometrical optics.)

The interesting question is whether, for Berkeley, a light ray is more than a theoretical entity, and is, in fact, a "hypothetical entity" that though useful to assume for the purpose of explanation in optics, has no real existence. This appears to be the view of Warnock [8] who would link "light" (or light rays) with "particles, atoms and the like" which, he says:

are not used in order to describe any observed facts, but in stating the general theory by which it is sought to account for those facts.

Warnock immediately goes on to quote a passage from *De Motu*:

They serve the purpose of mechanical science and reckoning: but to be of service to reckoning and mathematical demonstrations is one thing, to set forth the nature of things is another. (*De Motu* 18) [9]

However, Warnock quotes somewhat out of context, for the "they" referred to by Berkeley is not insensible particles, or atoms, but rather "forces" and particularly the technique of resolving "direct forces into oblique ones by means of the diagonal and sides of the parallelogram." (*De Motu* 18) The simplest interpretation of the passage is that the mathematical technique used in the resolution of direct forces does not commit one to assuming the real existence of those "oblique" forces represented by the "sides" of the parallelogram. There is nothing here that suggests Berkeley would consider as "hypothetical" entities the insensible particles of atomic physics or of physical optics.[10]

[8] Warnock, *op. cit.*, p. 201.
[9] *Ibid.*
[10] Warnock suggests that evidence for this view can be found in Berkeley's *Visual Language,* but we find no evidence there that Berkeley considers either light rays or their "insensible constituents" to be merely hypothetical entities. In section 37 of this work Berkeley writes: "A treatise, therefore, of this philosophical kind, for the understanding of vision, is at least as necessary as the physical consideration of the eye, nerve, coats, humors, refractions, bodily nature and motion of light, or the geometrical application of lines and angles for praxis or theory in dioptric glasses and mirrors, for computing and reducing to some rule and measure our judgments, so far as they are proportional to the objects of geometry. In these three lights vision should be considered, in order to be a complete theory of optics." Though the passage distinguishes geometrical from physical optics, there is no suggestion that hypotheses about the nature of light refer to nothing in reality. Moreover, passages in the *Siris,* (we will have occasion to refer to these) concerning Optics, clearly suggest that Berkeley accepts the view that light is composed of "insensible particles."

Our own reading of the *Essay* suggests that although Berkeley questioned the existence of the "lines and angles" of geometers, he does not appear to question the real existence of light rays. Berkeley's critical strictures in the *Essay* are against the hypostatization of mathematical entities ("lines and angles") into real beings, and it is this that links the *Essay* with the criticism of the concept of "force" found in *De Motu*. For example, in the *Essay*, when discussing monocular apprehension of distance, Berkeley remarks:

This confused appearance of the object does therefore seem to be the medium whereby the mind judges of distance in those cases wherein the most approved writers of optics will have it judge by the different divergency with which the rays flowing from the radiating point fall on the pupil. No man, I believe, will pretend to see or feel those imaginary angles that the rays are supposed to form according to their various inclinations on his eye. (*Essay* 22)

It is the "angles," not the "rays of light" that are claimed to be "imaginary."

Berkeley's critique of the hypostatization of mathematical objects, however sound in other places, appears misplaced here. In claiming that light travels in straight lines "rays" nothing is implied about the existence of lines and angles in themselves; those entities supposedly represented by the pencil lines and chalk marks of geometric opticians as they construct diagrams to help compute shadow lengths, refractive indexes, etc. Although there are important philosophic problems concerning the status of the objects of "pure geometry," the issue here is the status of light rays. And since light rays are not observable (as opposed to illuminated surfaces), Berkeley's evident acceptance of their existence does appear paradoxical, particularly in light of his general view that, except for minds, it is a necessary condition for (x) to exist, that (x) be perceived.

Yet the *Essay* (and the later *Visual Language*) offer impressive evidence that Berkeley in explaining certain visual phenomena, accepts both that light is propagated in straight lines, and that it is composed of certain "insensible" particles. The former, in conjunction with some physiological considerations concerning the eye, allows Berkeley to explain how, though we do not "see" by geometry, geometrical considerations are causally relevant in our metric judgments about the distance of certain objects. For example, the following passage in the *Essay*:

Confused vision is when the rays proceeding from each distinct point of the object are not accurately recollected in one corresponding point on the retina, but take up some space thereon – so that rays from different points become mixed and confused together. (*Essay* 35)

As a modern writer on optics expresses it; a "circle of confusion" is
formed on the retina, since the rays of light converge towards a point not
on the retina, but behind it.[11] There are two fundamentally distinct
aspects to this explanation of "confused vision." (A) There is a necessary
connection between the size of the retinal circle ("the circle of confusion") and the distance from the eye to the object. (Assuming here as
given a constant focusing power of the eye, and no angular distortion of
the rays between the object and the eye.) More explicitly, assuming both
the rectilinear propagation of light, and that distance is measured by a
straight edge assumed to obey the Euclidean axioms, it follows logically
that the closer the object, the larger the "circle of confusion." (B) There
is a strict correlation between the size of the "circle of confusion" and the
confusion in the visual field. Berkeley, unfortunately, is not explicit about
this correlation; his own comments at times suggest that he identifies the
retinal image with what is "seen." [12] For example, he remarks in section
38:

Thus the mind judging of the distance of an object by the confusedness of
its appearance, and this confusedness being greater or lesser to the naked eye

[11] M. H. Pirenne, "Physiological Mechanisms in the Perception of Distance, by
Sight and Berkeley's Theory of Vision," *British Journal Philosophy of Science,* 4
(1953-54): 13-21. The full passage (p. 14) is as follows: "If the rays come from the
point of an object very close to the eye, they diverge to such an extent that the normal
or hypermetropic eye cannot bring them to an accurate focus on the retina. That is,
even if the optical sstem of the eye is strong enough to render these rays convergent,
they will converge towards a point situated, not on the retina, but behind it. The
rays thus constitute a cone which intersects the retina, forming upon it a "circle of
confusion." Pirenne appears to agree with Berkeley that a myopic person would
associate greater confusion with greater distance. Pirenne contends that the physiological "accomodation" of the retina of the myope is an unlearned response; that, in
other words, the eye accomodates to confused vision by increasing or decreasing
the curvature of the lense in response not only to the *size* of the retinal circle of
confusion, but as well to the *angle* at which the rays hit the retina; and that this
response is independent of experience. This is, however, consistent with the view
that the myope would associate confused vision with greater distance.
[12] Berkeley's tendency to identify the retinal image with what is "seen," leads to
an unnecessarily complex solution of the problem of the "inverted image." In fact
there appear to be two distinct solutions to the problem. Solution (A) claims that it
is not legitimate to compare the "inverted" reinal imaget of a tree with the "top"
and "bottom" of the real tree, since the latter determinations are made on the basis
of tactual experience, and there is no necessary connection (in fact no way of comparing the heterogeneous visual and tactual space) between the visually and the
tactually identified "top," and "bottom." It is not clear why this is relevant unless
the retinal image is to be identified with what is "seen." The paradox concerns the
relative inversion of two visually identified items, my visual sense datum of a tree
and the visually identified retinal image of the tree. If we photograph both the tree
and the retinal image and place them side by side, the inversion becomes clear.
Solution (B) which seems more appropriate, points out that the internal relations
of the retinal image are exactly the same as those of the visual datum; the "retinal"
"roots" are near the retinal "earth," the visually identified "roots" are near the
visually identified "earth" and so forth. This structural isomorphism is sufficient to
show there is no "inversion" in any significant sense.

according as the object is seen by rays more or less diverging, it follows that a man may make use of the divergency of the rays in computing the apparent distance, though not for its own sake, yet on account of the confusion with which it is connected. But so it is, the confusion itself is entirely neglected by the mathematicians as having no necessary relation with distance, such as the greater or lesser angles of divergency are conceived to have. (*Essay* 38)

It is not clear whether "confusion" here refers to the "circle of confusion" on the retina, which strictly speaking is not seen, or the phenomenal confusion in the visual field. It is plausible that mathematicians would ignore the former in accounting for metric judgments of distance, since it is logically entailed by the degree of convergency or divergency of the rays. However, the judgments Berkeley is speaking of are those based on the continued association in the "mind" between distance and phenomenal confusion. Between the latter and the retinal circle of confusion there is an invariant but non-necessary connection. Yet it is this invariant and contingent connection between a physiological and a mental event which would account for the usefulness of geometrical optics in designing, for example, corrective lenses, and is presupposed in Berkeley's claim that although lines and angles are not immediately made use of in our phenomenal judgments of distance (by some natural geometry) they are causally relevant in such judgments since they are necessarily connected with the retinal circle of confusion.

Berkeley emphasizes that the association or connection between amount of confusion in the visual image and apparent distance is a contingent one, based on an experienced association between the relative nearness of the object and experienced confusion in the visual image. He suggests, in fact, that the "ordinary course of nature" might well have been different, "the farther off an object," "the more confused it should appear." (*Essay* 26) Implicit, however, in his discussion of the "Barrovian" case is that although the connection between visual confusion and distance is a contingent one, it is not primitive; it can be explained in terms of the principles of geometrical optics conjoined with an assumption of the invariable association between the retinal "circle of confusion" and experienced visual confusion. The "ordinary course of nature" involves an assumed association between a physiological datum and a conscious experience. Although the sense of "explain" used here is not foreign to Berkeley, (deducing an event or law from more general principles) what might be problematic is the nature of this association between a bodily and a mental event [13] However, the issue is not raised here, primarily, we

[13] What we can call the problem of mind body interaction is dealt with more explicitly in the *Three Dialogues Between Hylas and Philonous,* in the context of

think, because Berkeley appears to identify the physiological datum (the retinal "circle of confusion") with the mental event. (The experience of confusion in the visual field.)

Berkeley makes use not only of hypotheses concerning the geometric properties of light, but appears to presuppose something about light's physical nature.[14] Discussing the fact that the moon at the horizon appears larger than the meridian, although the angle subtended by the eye is the same, Berkeley suggests that:

the particles which compose our atmosphere intercept the rays of light proceeding from any object to the eye; and by how much the greater is the portion of atmosphere interjacent between the object and the eye; by so much the more are the rays intercepted, and, by consequence, the appearance of the object rendered more faint, every object appearing more vigorous or more faint in proportion as it sends more or fewer rays into the eye. (*Essay* 68)

Not only does the passage indicate some view concerning the physical nature of light, (not here spelled out) but it suggests a "causal" relation between a physiological datum (number of light) rays reaching the eye) and a mental event ("faint" appearance of the object). Unlike the earlier case, however, the dependency is explicitly mentioned. Moreover we again have a particular contingent relation of association; here association between "faintness" and apparent magnitude itself explained (deducible from) certain physical properties of light conjoined with an assumption concerning an invariant association between the number or rays reaching the eye and the degree of "faintness" in the visual experience.

The *Essay* appears to abound, then, in materialist imagery. And it is

certain transmission theories of perception. We will discuss it in more detail, when we deal with the concept of "matter." We should note here, that although Berkeley allows for "psycho-physical" laws, it would be incorrect to construe these as invariable relations between kinds of "ideas," where "idea" has the meaning, "object of sense." In describing something called "experience of visual confusion," or the "experience of redness," the attribution of consciousness is an essential ingredient in the description of the phenomenon. This is not true if I wish merely to describe the visual field itself. From this point of view the following remark of Philonous in the Second Dialogue is misleading. (Turbayne ed. p. 150) "It comes, therefore, to the same thing; and you have been all this while accounting for ideas by certain motions or impressions in the brain, that is, by some alterations in an idea, whether sensible or imaginable it matters not." Again, construing "idea" as "object of sense" what is being accounted for by establishing correlations between states of the sensible brain and mental states (like "seeing red") cannot strictly called "ideas."

[14] Berkeley was evidently aware of reflections on the physical nature of light. In the *Siris*, extensive use is made of Newton's *Optics*. On the Newtonian texts available to Berkeley at various stages of the latter's career, see: John W. Davis, "Berkeley, Newton and Space," in, R. Butts and J. W. Davis ed., *The Methodological Heritage of Newton*, (London; Basil Blackwell; 1970) 57-74.

just such considerations that led Pirenne, whom we have previously quoted, to comment:

Berkeley seems to place man in an external world, the extension of which he can know by touch. Light stimuli excite his retina, leading to sensations and perceptions in his mind.[15]

The later *Visual Language* attempts to be more specific about the nature of "light rays," here calling them "tangible objects." The context is a discussion of the inverted retinal image, and the full passage is worth quoting:

The pictures, so called, (on the retina) being formed by the radius pencils after their above mentioned crossing and refraction are not so truly pictures as images or figures or projections, tangible figures projected by tangible rays on a tangible retina, which are so far from being the proper objects of sight that they are not at all perceived thereby, being by nature altogether of the tangible kind and apprehended by the imagination alone, when we suppose them actually taken in by the eye. (*Visual Language* 49)

Unlike the *Essay*, a clear distinction is made here between the retinal image (which is called a "tangible object") and the content of purely visual experience (the "proper objects of sight"). Unfortunately, however, there is an ambiguity in the concept of a "tangible object," an ambiguity essentially related to the ambiguity in the concept of distance. On the one hand "tangible object" could mean those objects in real three dimensional space whose essential nature is revealed to touch. Berkeley will admit, in fact, that this is the way he expects the *Essay* to be understood, although protesting that his locutions are a mere *façon de parler*, a "vulgar error," as he says in the *Principles*. (*Principles* 44). On the other hand "tangible object" could refer to tactual sense data (the "proper objects of touch") where, however, "tactual" refers both to cutaneous and kinaesthetic sense data. There is a clear parallel between these two definitions of "tangible object" and the two senses of "distance" previously mentioned; distance as (1) a metric property, the judgment of which can be the result of applied geometry, and (2) distance as an apparent quality of the visual field itself. (The apprehension of "outness.")

There is, however, more here than a mere parallel: there is an essential relation between the senses of these two concepts. One way of elucidating

[15] Pirenne, *op. cit.*, *p.* 20. Also see "The Moon Illusion," Lloyd Kaufman and Irving Rock, *Scientific American*, July 1962, Vol. 207, No. 1, pp. 120-132. Authors conclude: that changes in the relative brightness of the horizontal and zenith moon have no effect on the illusion. (p. 128)

this is to notice that Berkeley appears to draw back from claiming that "light rays" and "retinal images" can literally be touched:

And here it may not be amiss to observe that figures and motions which cannot actually be felt by us, but only imagined, may nevertheless be esteemed tangible ideas, forasmuch as they are of the same kind with the objects of touch, and as the imagination drew them from that sense. (*Visual Lang.* 51)

On the face of it, this is a baffling passage. What after all, are the essential properties of "light rays" and "retinal images," that although not revealed directly to touch, are analogous to something we are (according to Berkeley) aware of through tactual experience? It seems clear that what is referred to here is the metric property of length and area. The "pencils of (light) rays" (that might be signified by the lines of the physicist's diagrams) are grasped by the "imagination" as analogous to the straight edges of physical objects. The "figure" and "magnitude" of the retinal image are properties that would be ascribed to it by either direct measurement or through the application of Euclidean geometry. It is because the light rays and retinal image share the same geometrical structure as the "radiating surface," that Berkeley can say . . . "those images (retinal images) when the distance is given should be simply as (directly proportional to) the radiating surfaces. (*Visual Language* 53)

Berkeley goes on to say that although the "pictures" or proper objects of sight are heterogeneous with respect to the "images" (what is on the retina), the former "may be proportional" to the latter. Therefore as a matter of contingent fact, the "pictures" (content of the visual field) may be "proportional to those radiating surfaces or the *real tangible magnitude of things.*" (*Visual Language* 53) We are not concerned here with Berkeley's argument that geometrical and physiological considerations show that there is often a proportionality between apparent (phenomenal) and real (metric) magnitude, although no innate geometry allows us to compute the latter a priori.[16] What is of immediate

---

[16] Berkeley does distinguish here between the retinal image and the visual experience. The former is called a "tangible object." However, no reason is offered for the proportionality between the magnitude of the image of an object on the retina and the visual experience of the magnitude of the object. However, it is this mediating and contingent correlation between the magnitude of the retinal image and judgments of magnitude, that explains the proportionality between "real" magnitude and judgments of magnitude. If (A) represents "real magnitude," and (B) apparent or judged magnitude, and (C) the magnitude of the retinal image of the object Berkeley's argument is as follows. Assuming the rectilinear propagation of light, (C) can be deduced from knowledge of (A). (C) and (B) are invariably but contingently related. (Has the form of a psycho-physical law.) Therefore the proportionality between (A) and (B) can be deduced ("explained"). Although the truth of the proportionality between (A) and (B) is logically entailed by the truth of Euclidean

interest is his reference to the "radiating surfaces" of physical objects as "the real tangible magnitude of things." This suggests that for Berkeley the metric properties of objects – those that enter into the computations of applied geometry and more generally into physical laws – have an essential relation to tactual experience. There is the germ here, along with complementary passages in the *Essay*, of a theory of measurement, and we will in the next section discuss Berkeley's notions concerning measurement in more detail. We can, however, briefly mention what, in his view is the problem concerning measurement. Visual judgments of magnitude, he will claim, are not invariant with respect to the visual perspective of the observer. Yet scientific laws require such perspective invariant judgments of the magnitude of objects Since vision is not the source of such judgments, and there do exist mathematically formulated laws, the source for such judgments must be another sense, namely touch.

Two last points should be mentioned before completing this section. The first is, that although Berkeley more carefully distinguishes, in *Visual Language*, the "retinal image" as a tangible object, from the content of visual experience, he is not very successful in his brief remarks on the nature of light. His suggestion that the "ray" like character of light is grasped by the "imagination" is not much help. We find no evidence that, as Warnock suggests,[17] Berkeley took "light ray" to be a hypothetical

geometry conjoined with a psycho-physical law, it is a contingent truth (synthetic). Certain givens are assumed here; for example, no angular distortion of the light rays by a lense between the eye and the object, no atmospheric distortion, etc. Changing these givens, however, does not alter the structure of the argument; (C) is still logically entailed by (A) conjoined with information about angular distortion of the rays; (C) is still considered to be contingently but invariably associated with (B): in the Barrovian case, for example, (*Essay* 29) a retinal "circle of confusion," is still associated with visual experience of confusion and a judgment of "nearness." In all cases the propositions of geometry (with "light ray" identified as "straight line") plus certain psycho-physical laws "explain" the relation between (A) and (B).

In Newton's *Optics,* there is a strong identification of the retinal image with what is "seen." Newton speaks of the "pictures" (retinal images) "propagated by Motion along the Fibres of the Optic Nerves into the Brain," which is "the cause of vision." From this point of view it becomes easy to assert that "accordingly as the pictures are perfect or imperfect, the object is seen perfectly or imperfectly." (Isaac Newton, *Optics,* (New York: Dover Publications, 1952; based on the Fourth Edition, London; 1730) p. 15.)

Certainly by the time of *Visual Language,* Berkeley has clearly distinguished the retinal image from the content of the visual experience, but maintains there is a structural isomorphism between the "image" and the visual field.

[17] Warnock *op. cit.,* p. 203. One could perhaps refer to a passage in the *Dialogues Between Hylas and Philonous* (Dialogue 1, Turbayne ed. p. 125) as evidence that Berkeley would deny the existence of the particulate nature of light. Hylas remarks that "a distant object, therefore, cannot act on the eye, nor consequently make itself or its properties perceivable to the soul. Whence it plainly follows that it is immediately some continguous substance which, operating on the eye, occasions a perception of colors; and such is light." Philonous (Berkeley) then remarks somewhat incredulously. "How! is light then a substance?" Hylas then says: "External light is nothing but a

entity or "mathematical fiction," as he later understands the term "force" in mechanics. The second point is that *Visual Language* like the earlier *Essay*, appears to place man, as Pirenne suggests, in the "Cartesian Space." [18] Berkeley speaks of "radiating surface," light rays, sense organs (eyes) and their parts (retina), suggesting a world of material objects that can through the transmission of impulse (radiating surfaces produce light rays which interact with the retina) produce sensations. In part the problem concerns what Berkeley meant by "tangible object"; whether the term refers to "objects" that exist independently from minds but whose essential structure (particularly magnitude) is revealed to touch; or whether "tangible object" is synonymous with a certain congeries of tactual sensations. Specifically our problem will be whether such metric properties as length, area, weight, etc., are in any essential way related to tactual experience.

THE VULGAR ERROR

Strictly speaking the phrase "vulgar error" refers to Berkeley's contention in the *Principles* (44) that the apparent assumption in the *Essay* of a world of material objects existing independently of perception, was a mere facon de parler.

That the proper objects of sight neither exist without the mind, nor are the images of external things, was shown even in that treatise. (*Essay*)

thin fluid substance whose minute particles, being agitated with a brisk motion and in various manners reflected from the different surfaces of outward objects to the eyes, communicate different motions to the optic nerves; which being propagated to the brain, cause therein various impressions, and these are attended with the sensations of red, blue, yellow, etc." It turns out, however, that the question of the existence of "insensible particles" of light is not under discussion: rather what Philonous objects to here, is the predication of color terms ("red," "yellow," etc.) of the unobserved particles (or "motions" of the particles) that allegedly constitute the physical nature of light.

In the *Siris* there is some extended and speculative discussion of the physical nature of light, modeled after Newton's "queries" appended to his *Optics*. In section 155, Berkeley distinguishes the "aether, fire, or substance of light," as the primary instrumental cause, from "mind" which is the active or efficient cause. In section 169 he remarks: "This pure fire, aether or substance of light was accounted in itself invisible and imperceptible to all the senses, being perceived only by its effects, such as heat, flame and rarefaction . . ." And section 162: "The pure aether or invisible fire (which Berkeley identifies with light) contains parts of different kinds that are impressed with different forces, or subjected to different laws of motion, attraction, repulsion and expansion, and endued with divers distinct habitudes towards other bodies." We will discuss these passages and others, when we deal with Berkeley's view of "corpuscularism," noting now only that if the *Siris* is taken seriously as a speculative work in science, it is difficult not to come away with the conclusion that Berkeley accepts that light is ultimately composed of insensible particles.

[18] Pirenne, *loc. cit.*

Though throughout the same, the contrary be supposed true of tangible objects – not to suppose that vulgar error was necessary for establishing the notion therein laid down, but because it was beside my purpose to examine and refute it in a discourse concerning vision.

Commentators vary in how seriously they take the "error" to be in the works on vision, and as to how successful is Berkeley's disavowal in the *Principles*.[19] We will not deal with all the aspects of this issue, but rather attempt to focus on those aspects that most directly relate to the problem of measurement; particularly a distinction Berkeley attempts to draw between a purely visual metric and a metric ascertained by geometry and somehow essentially related to tactual experience.

---

[19] Norman Kemp Smith (*Commentary to Kant's Critique of Pure Reason* p. 588) argues that Berkeley's immaterialism presupposes the existence of material sense organs. Smith's view discussed by John Wild, *George Berkeley*, (New York: Russell and Russell, Inc., 1962-first published 1936, Harvard University Press.) P. 91. Wild believes Smith to be mistaken, and that Berkeley's mature position would require no reference to "physiology or anatomy." (Wild, p. 91.) Turbayne (Introduction and Commentary to the Works on Vision, *op. cit.*) appears to find Berkeley's disclaimer in the *Principles* satisfactory, and in our judgment, ignores the importance of materialist arguments in both the *Essay on Vision,* and the later *Visual Language.*

Wild (*op. cit.*, pp 89-92) appears to have the more balanced view. He recognized that: (A) the use of materialist arguments in the *Essay,* and *Visual Language,* were more than a manner of speaking: (B) that Berkeley was attempting to free himself from "physiological" presuppositions and construct a phenomenology (what Berkeley calls a "philosophy") of vision. Often Berkeley reminds us that vision is properly attributed to the "soul" and not the "eye." (C) the programmatic nature of this "philosophy" of vision. Berkeley never carries out in great detail the correlations between strictly visual (light and colors) tactual and kineasthetic experiences that would result in the sense of "spatiality." The passage in the *Principles* (44) where the "vulgar error" is mentioned suggests such a program. "So that in strict truth the ideas of sight, when we apprehend by them distance, and things placed at a distance, do not suggest to us or mark out to us things actually placed at a distance, but only admonish us what ideas of touch will be imprinted on our minds at such and such distances of time, and in consequences of such and such actions." Wild admits that an essential stumbling block in the program is the concept of kinaesthetic sensations as involved in the genesis of our sense of spatiality, since the sense of space (Berkeley will admit this in *De Motu*) is involved in the "idea" of motion.

Warnock (*op. cit.*, Ch. 3) recognizes that the *Principles* is suggesting a more radical phenomenalism in the sense that the expression "object of touch" or "tangible object" will no longer refer to objects that exist in a three dimensional space external to and independent of the perceiver, objects whose "real" nature is somehow disclosed to touch; but will refer to a certain congeries of tactual sensations, which like the objects of any other sense have no existence external to consciousness. The problem with such a translation of "tangible object" is that it becomes unintelligible why correlations between visual and tactual sensations should ever result in the sense of "spatiality" or "outness" that we allege is a characteristic of the visual field.

D. M. Armstrong (*Berkeley's Theory of Vision,* (Melbourn: Melbourn University Press; 1961) contends that Berkeley merely uncritically accepted an assumption from previous optical theory (that "distance" is not immediately seen) and that he identifies what is "seen" with the simulacrum that appears on the "fund" (retina) of the eye. Since the simulacrum is two dimensional, what is immediately seen then must be two dimensional. In Visual Language, however, Berkeley does distinguish between the retinal image and the visual object, but never quite frees himself from the belief that the dimensional structure of the "proper" visual object must be identical with that of the retinal image.

One could contend that the "error" appears at the very beginning of the *Essay*, when Berkeley, placing himself in agreement with Molyneux and others,[20] contends that "distance of itself and immediateley, cannot be seen." "For," Berkeley continues:

distance being a line directed endwise to the eye, it projects only one point in the fund of the eye, which point remains invariably the same, whether the distance be longer or shorter. (*Essay* 2)

On the surface the argument appears to assume (A) the existence of material objects at various distances from an observer, and (B) an identity, or at least dimensional isomorphism between the "retinal image" and what is seen. Assumption (A) appears to require the existence of a real space in which the observer is placed; a view Berkeley believes refuted by the conclusions of the *Essay* itself. Assumption (B) is rejected in *Visual Language,* at least in so far as the Former claims that "visual object" and "retinal image" are identical in reference The work does suggest, however, that there is some structural isomorphism ("proportionality" is Berkeley's term) between the retinal image and the object of vision. Thus it is difficult not to conclude that there is either a clear inconsistency or at least arbitrariness in this contention that "distance" cannot be immediately seen. If the Molyneux premise is taken seriously, then Berkeley's argument presupposes the existence of that which he is later at pains to deny; that is, three dimensional space. If the Molyneux premise is not taken seriously it no longer logically functions to support the claim that distance cannot be immediately seen. We are not concerned at this point with Berkeley's view that this claim is also an empirical one, for which evidence concerning what is seen by the "born blind" who are given vision, is relevant.[21]

One might defend Berkeley by saying that his aim is not to question the premise of Molyneux,[22] but that he agrees that "distance is not im-

---

[20] Molyneux, *New Dioptrics* (London: 1692) p. 113. Quoted by Turbayne, *Berkeley's Works on Vision, op. cit.,* p. 19. "Distance of itself is not to be perceived. For it is a line presented to our eyes with its end toward us, which must therefore be only a point, and this is invisible."

[21] A rather detailed investigation of the evidence related to the Molyneux problem is found in John W. Davis, "The Molyneux Problem," *Journal of the History of Ideas:* Vol. 21 No. 3, July-Sept. 1960, pp. 392-408. There is also consideration of the problem in some detail by Berkeley's early critics, Samuel Bailey and Thomas Abbott, neither of which is mentioned by Davis. See Samuel Bailey, *A Review of Berkeley's Theory of Vision* (London: 1842) and Thomas K. Abbott, *Sight and Touch* (London: 1864).

[22] In this regard, it is difficult to understand Turbayne's contention (*Berkeley's Works on Vision op. cit.,* p. 19 footnote) that the "view accepted both by Berkeley

mediately seen," yet claims nevertheless judgments of distance are not made through use of an innate geometry, but rather through complex associations of "properly" visual (light and colors) tactual and kinaesthetic experiences. Such a defense, however, in identifying the "proper object of sight" as a two dimensional manifold of light and colors, begs the question as to whether "distance" is a "proper object of sight."

In fact, if the Molyneux premise is rejected, on what grounds would one reject the view that the sense of "outness" or spatiality is constitutive of all visual consciousness? [23] One might respond that Berkeley is not speaking of this consciousness of spatiality, but rather of metric judgment of distance. Wild appears to take this position. He contends that Berkeley's most "telling" argument for the view that tri-dimensionality is not a proper object of sight, is not physiological arguments concerning the retina, nor evidence of what is perceived by the born blind who have vision restored, but

on the fact that the visual magnitude of objects is not what we mean by the "actual" size of an object, – the moon for example. The visual size or magnitude may vary, while the "actual" size and position remain the same. *This of course implies that visual objects have some size, though not the actual size.* (Our italics.) What we mean by the actual size, distance and situation of objects is not something that is directly given in the experience of seeing.[24]

The distinction between perspective relative "visual magnitude" and invariant "tangible" (or "actual") magnitude, is an important one for Berkeley, and Wild correctly points to it. We will later consider it in some detail. However, in attempting to tidy up Berkeley's use of the term "distance," Wild distorts the implication of the texts. The latter often make it clear that Berkeley is referring to the experienced sense of "outness" or spatiality when he denies that "distance" can be a proper object of sight. The burden of evidence from cases of the born blind who have restored vision is to show precisely that. For example, in the *Essay* after thoroughly criticizing the claim of previous writers on optics that we judge distance by an innate or "natural" geometry, and that our judg-

and his opponents, that distance is not immediately seen is a conclusion, not a premiss of this theory of vision." Berkeley certainly appears to accept it as a premiss. It is true that he apparently felt certain empirical considerations relating to the visual field of the born blind who regained sight, were sufficient evidence for his claim, but he never claims such considerations were necessary since "a priori" considerations were equally sufficient.

[23] For a defense of the thesis that "outness" or "spatiality" is constitutive of all visual consciousness, see William James, *The Principles of Psychology* (Vol. 2, original copyright 1890, Dover reprint 1950, pp. 212-215).

[24] Wild, *op. cit.*, p. 93.

ments of distance are, in fact, dependent on certain perceptual cues such
as "strain," "faintness" or "confusion," Berkeley remarks:

> From what has been premissed, it is a manifest consequence that a man born
> blind, being made to see, would at first have no idea of distance by sight:
> the sun and stars, the remotest objects as well as the nearer, would all seem
> to be in his eye, or rather in his mind. The objects intromitted by sight would
> seem to him (as in truth they are) no other than a new set of thought or
> sensations, each whereof is as near to him as the perceptions of pain or
> pleasure, or the most inward passions of his soul. For our judging objects
> perceived by sight to be at any distance, or without the mind is entirely the
> effect of experience, which one in those circumstances could not yet have
> attained to. (*Essay* 41)

Berkeley did not, in our judgment, sufficiently distinguish the visual
consciousness of space from metric judgments of distance and magnitude.
On the other hand, the *Essay* makes it quite clear that even the apparent
visual consciousness of space is not for him a primitive visual datum, but
a resultant generated by the association of "properly" visual sensations
with tactual (including kinaesthetic) experiences. What we call our
awareness of space is, for Berkeley, a set of implicit metric judgments
More precisely, certain metric judgments become associated with certain
visual ("faintness," "confusion") or organic ("strain") experiences, and
then as it were animate later occurrences of these experiences giving us
the sense of space, much as the meanings of words come to animate the
particular written or spoken vehicles, giving us the sense of significant
discourse.

Yet, however, valid the above view, particularly the analogy drawn
between our awareness of space, and our awareness of meaningful discourse,[25] the Molyneux premiss is not relevant. It is difficult to escape

[25] See Bailey, *op. cit.,* pp. 86-87. Berkeley's use of the analogy is in the *Essay*
(sec. 17). "... there has grown a habitual or customary connection between these
two sorts of ideas so that the mind no sooner perceives the sensation arising from the
different turn it gives the eyes, in order to bring the pupils nearer or farther asunder,
but it withal perceives the different idea of distance which was wont to be connected
with that sensation. Just as, upon hearing a certain sound, the idea is immediately
suggested to the understanding which custom had united with it." There are numerous
difficulties with this position. (A) As we have mentioned this view appears to run
counter to the position of the *Alciphron* and the *Principles,* that it is not necessary
that the sensible referent for a term need be present (qua image) in order for the
term to be understood. (B) Even if the view of understanding language presented here
is correct, the analogy appears to break down with respect to the visual perception of
"distance." The point is made by Bailey (*op. cit.,* p. 88). He writes: "When the word
'moon' is pronounced in my hearing, there is an idea of image of the planet raised
up in my mind; but when I see the distance of the chair, I am unconscious of anything
but the simplest sight of the piece of furniture and of the interjacent ground. No idea,
no copy of any tactual perception of distance presents itself... This appeal to
consciousness appears to me, I confess, quite conclusive against the theory of Ber-

the conclusion that Berkeley accepts the premise of Molyneux because he assumes if not an identity of reference for "retinal image" and "visual object," at least a structural isomorphism between their referents. Thomas Abbott, among Berkeley's critics, suggested that the assumption of such an isomorphism accounted for the latter's acceptance of the Molyneux premise.

The fact is that the argument (Molyneux's argument that distance cannot be immediately seen) – implies that the antecedent to the perception of distance must be a sensible interval in the organ, in other words that the organic affection must resemble the perception. But as just remarked, we can assume no such a priori conditions.[26]

Abbott would seem to be correct; the fact that "distance" is a line directed endwise to the eye, is no evidence at all that "outness," or spatiality is not a proper object of sight. Berkeley may have felt that to admit spatiality as a proper visual datum would imply the existence of space independent of the perceiver. However, there is no such implication; whether spatiality (the appearance of objects to a perceiver as a three dimensional manifold) is or is not intrinsic to all visual consciousness, has no bearing on the issue of the ideality of space.[27]

Although Wild's interpretation somewhat constricts Berkeley's thesis, and tends to ignore some crucial passages, it does direct us to look more critically at the concept of distance as it is used in the *Essay*. Berkeley continually posits an experienced association between "distance" and sensations such as visual confusion in the visual field, faintness of the image, etc. For example, consider section 3 of the *Essay:*

Again, when an object appears faint and small which at a *near distance*

keley." Bailey's view is that the referent for the term "distance" in Berkeley's theory is some tactual sense datum, and that in apprehending distance we do not evoke in imagination any such tactual "idea." Nor do we call up any "kinaesthetic" sensation, and image of any movement of my body towards the object. This is consistent, we might add, with the view that tactual experiences may be relevant in the genesis of our sense of space. As we will see, there is in Berkeley's discussion a serious equivocation in his concept of "distance"; on the one hand it refers to our allegedly visual experience of "outness" or "spatiality"; on the other hand it refers to metric judgments of distance.

[26] Abbott, *op. cit.*, p. 12.

[27] The point is made by Armstrong. Even if we consider "tri-dimensionality" to be constitutive of all visual experience, this would not establish that space is transcendentally real; that is, existing independently of any perceiver. In the *Three Dialogues Between Hylas and Philonous,* Berkeley evidently recognizes this point himself. In the first dialogue (Turbayne ed. *op. cit.*, p. 143) Philonous contends: "But allowing that distance was truly and immediately perceived by the mind, yet it would not thence follow it existed out of the mind. For whatever is immediately perceived is an idea; and can any *idea* exist out of the mind?" (Berkeley's emphasis.) The *Essay* then, for Berkeley, should have no bearing on the truth of immaterialism.

I have experienced to make a vigorous and large appearance, I instantly conclude it to be far off.

Or section 16:

And first, it is certain by experience that when we look at a *near* object with both eyes, according as it approaches or recedes from us, we alter the disposition of our eyes by lessening or widening the interval between the pupils. This disposition or turn of the eyes is attended with a sensation which seems to me to be that which in this case brings the idea of *greater* or *lesser distance* into the mind.

And finally section 25:

Thus greater confusion having been constantly attended with *nearer distance,* no sooner is the former "idea" perceived, but it suggests the latter to our thought. (My italics.)

Examples could be multiplied, without, however, altering the argument. The question in all cases of an alleged association between "distance" and visual confusion or "distance" and faintness of image etc., is the meaning or referent of the term "distance." Although Berkeley writes (section 16) of "bringing the idea of a greater or lesser distance into the mind," the language is loose, for "distance" is not for him an "idea," if we think of "idea" as the proper object of a sense. Berkeley appears to admit as much and draws an interesting analogy between our alleged perception of distance and recognition (visual) of another person's emotions.

As we see distance, so we see magnitude. And we see both in the same way that we see shame or anger in the looks of a man. Those passions are themselves invisible; they are nevertheless let in by the eye along with colors and alterations of countenance which are the immediate object of vision, and which signify them for no other reason than barely because they have been observed to accompany them. Without which experience we no more should have taken blushing for a sign of shame than of gladness. (*Essay* 65)

Strictly speaking, the analogy is drawn to demonstrate that "distance" is not a proper object of sight, leaving it open whether it is a proper object of some other sense. However, the analogy suggests that it is not a proper object of any other sense, for like "shame" or "rage" the claim is that it is "seen," albeit mediately. As Berkeley's early critics pointed out, this puts "distance" in rather a unique category The musician when reading a score, although he has in imagination something structurally analogous to hearing, does not believe he is "seeing" the music. Bailey makes the point in general.

It may be laid down as a general law, that when the mind neglects the sign and passes onto the thing signified ... it is apt to fancy not, as Berkeley's theory requires, that it perceives the suggested objects with the sense actually in exercise, but that it perceives them with the sense to which the conceptions belong, or from which they have been derived.[28]

Yet we apparently "see" distance; and this tri-dimensional quality of the perceptual field results, according to Berkeley, from continued association of certain sensations like visual strain with something called "distance." We are still faced with the question of to what the latter refers, as, referring back to the analogy, we are faced with the question of the referent for "shame" or "rage."

The simplest answer is that distance terms (e.q. "nearness") used in the above associations refer to metric distance; distance as directly measured, or indirectly measured through the application of Euclidean geometry. The same would hold true for "magnitude" in associations, for example, between faintness of image and the magnitude of objects, the latter referring to what Berkeley calls in *Visual Language,* the "real tangible magnitude" of things. (*Visual Language* 53.) This answer is plausible, we believe, not only from an analysis of the texts, but because of the following formal consideration. Although we apparently "see" distance and magnitude, such perception is, for Berkeley, an implicit judgment concerning the distance and magnitude of objects. It is an implicit judgment in the same sense that for Berkeley my perception of a man in the street would be an implicit judgment. By implicit judgments we understand claims ("X is a man"; "I hear a coach"; "that rock is smooth," "The moon is 60 semi-diameter's of the earth distant from me") that in principle refer to certain experiences which are not yet occurring. Berkeley gives us an excellent example in *Principles* 44.

So that in strict truth the ideas of sight, when we apprehend by them distance and things placed at a distance, do not suggest or mark out to us things actually existing at a distance, but only admonish us what ideas of touch will be imprinted in our minds at such and such distances of time, and in consequence of such and such actions.

For example, my visual apprehension of a rock as "smooth" is in part a claim that as a consequence of certain actions of my hand, I will get certain tactual sensations. Berkeley's interesting claims, however, do not concern clear tactual qualities like "roughness" or "smoothness," but qualities such as distance and magnitude. (And figure) And it is here where there are difficulties in understanding his position. For example,

[28] Bailey *op. cit.,* p. 87.

my apparently visual perception that (x) is round, might according to
this theory mean that upon certain movements of my hand (tactually
tracing the boundary of the object), I will get certain sensations of touch.
This appears false on the surface, since often I cannot discriminate
shapes by touch (an ellipse from a circle) that I can by sight.[29] Yet
Berkeley might have rejoined that my visual judgments of shape are
relative to my visual perspective and therefore useless for scientific work.
(If I wish to compute the volume of a solid using the area of a right
section and the altitude, for example.) Thus "roundness" may mean for
him the geometric property of roundness, ascertained for example by
some process of measurement. In the case of "distance," the phrase
"such and such actions" might again refer to a process of measurement
and our problem will be to understand a relation Berkeley is evidently
attempting to establish between measurement and tactual experience.
Thus, in the experienced associations between "distance" and sensations,
the former does not mean for Berkeley (a) the "quality" of tri-dimensionality in the visual field, since that is a *result* of such associations; nor is
it plausible, we believe, to construe it as (b) some kind of raw tactual
feel comparable to "roughness" or "smoothness." The most plausible
alternative, then, is to construe "distance" and "magnitude" as metric
distance and magnitude.

Mathematically formulated laws of nature (for example in geometrical
optics) deal with the metric properties of objects; they require determinations of distance and magnitude, that according to Berkeley, are
invariant in two respects; (a) with respect to the visual perspective of an
individual observer, and (b) invariant for all observers (intersubjective
agreement). Although (b) raises important problems,[30] our observations

---

[29] Abbott (*op. cit.,* p. 33) offers this kind of evidence as an "experimentum crucis,"
against Berkeley. Abbott's point is that "touch" is a very poor sense for disclosing
shape; that we often check our tactual identifications visually. Moreover visual judgments of size difference (for example, of two coins held an equal distance from my
eyes) does not imply that there will be any recognizable difference in our tactual
experience; visual discriminations of figure are much finer than tactual ones. The
contention is that visual discriminations are often not signs or cues for any possible
tactual experience.

[30] The problem of (b) is generally discussed in terms of whether such required
invariancy (intersubjective agreement) is compatible with immaterialism. Gerard
Hinrich in a monograph has argued that the scientific requirement of intersubjective
agreement is not compatible with immaterialism. (See Gerard Hinrich, "Berkeley on
Size and a Common World," *Personalist,* 1951, pp. 251-258.) More specifically
Hinrich's point is that when Berkeley in the *Principles* and the *Three Dialogues,*
attempts to remove from the sense of "tangible object" any suggestion of reference to
objects existing independent of mind, and equate "tangible object" with certain tactual
sensations, he can no longer maintain the doctrine or size invariancy for "tangible
extension" as opposed to "visible extension." The point is well taken, and as we argue

here will concern (a). We find that Berkeley gives Locke's concept of primary qualities an operational cast; that it is the invariance of metric length that establishes it as a primary quality, not some ontological privilege of being a property of the object as opposed to a modification in the subject.[31]

in the text, that while Berkeley admirably indicates in the *Essay* a need for a concept of size invariancy he never makes it clear how this invariancy is related to the concept of "tangible magnitude" particularly if the latter means (as *Principles* 44 suggests it should) magnitude revealed as a tactual sense datum. However, against Hinrich, one could argue that the existence of a *mind-independent* "common" world is not required as a presupposition to make sense of spatial measurement. The selection of a "unit," the determination of equality of segments in terms of the observed coincidence of end points, and the assumption of invariancy over transport are all compatible with an immaterialist metaphysics.

[31] Berkeley's criticism of the alleged distinction between primary and secondary qualities is often on the grounds that the former like the latter fail to meet the invariancy conditions mentioned in (a) and (b) (our text p. 62). For example *Principles* (11), speaking of "motion." "Again great and small, swift and slow are allowed to exist nowhere without the mind, being entirely relative, and changing as the frame or position of the organs of sense varies." Yet in the *Essay,* Berkeley has in fact argued that "tangible extension" is size invariant with respect to the position of the observer. In the *Three Dialogues* (Dialogue 1, Turbayne ed. p. 128) there is the following comment of Philonous: "But as we approach to or recede from an object, the visible extension varies, being at one distance ten or a hundred times greater than at another..." Philonous goes on to suggest that just as "heat" or "cold" cannot be *in* the water because such qualities are not invariant with respect to the condition of the subject, so, "figure" cannot be in the object, since to one eye an object could look "smooth and round," and to the other "great, uneven and angular." (p. 129) What is interesting is that in neither case is there any reference to "tangible" magnitude or figure, which the *Essay on Vision* (and *Visual Language*) contends has the requisite invariance. It is passages like the above which Hinrich seizes upon to demonstrate his contention that if "size" is a "proper" sensible as opposed to a "common" sensible, and is thus relative to the nature of the perceiver (that is to the condition of the perceiver) there could be no general laws which presuppose inter-subjective agreement as to the size of objects. His view is that if size is a "proper" object of sense, in the sense that size judgments are essentially private (relative to the condition of the observer), that appeal to God as the source of the intersubjectively held "laws" of nature is not sufficient. The "roundness" that appears to one eye and the "angularity" that appears to the other equally originate with God; we are not speaking of a difference between percepts and images. The point is well taken; clearly reference to God as the source of percepts cannot in itself insure that there will be common laws, particularly if size, for example, *means* no more than a type of judgment relative to the visual perspective of the observer. Berkeley, of course, indicates as much in the *Essay*: that we require a concept of size which includes the required notion of invariance with respect both to the visual perspective of a single observer, and the various perspectives of a multiplicity of observers. Yet in the *Principles* and the *Three Dialogues,* where the emphasis is on destroying the distinction between primary and secondary qualities, size is presented in such a way that is becomes meaningless to speak of a real "size" of an object in the same sense that it is meaningless to speak of a real color of an object. What is needed is an explication of the sense in which we can have intersubjective agreement that we are speaking of the same object, and a method for establishing agreement as to the size of that object. As to the first, Berkeley suggests that although strictly speaking all percepts are private (possessed by individual minds) there is no objection to denominating some of these with the term "same." For example, Dialogue 3, (p. 194), Philonous: "Let us suppose several men together, all endowed with the same faculties,

The following passage clearly illustrates Berkeley's concern with the issue of measurement.

The magnitude of the object which exists without the mind, and is at a distance, continues always invariably the same; but the visible object still changing as you approach to or recede from the tangible object, it has no fixed or determinate greatness. Whenever, therefore, we speak of the magnitude of anything, for instance a tree or a house, we must mean the tangible magnitude; otherwise there can be nothing steady and free from ambiguity spoken of it. (*Essay* 55)

We can take this to mean that in order for our metric locutions concerning magnitude to be useful, they must refer to a property that remains the same, regardless of our distance from the object. Berkeley calls this "tangible magnitude" and opposes it to "visual magnitude," which does vary with our distance from the object. In a later passage, and again in reference to magnitude, Berkeley becomes more explicit about the methodological significance of this distinction between "visible" and "tangible" magnitude.

Inches, feet, etc., are settled stated lengths, whereby we measure objects and estimate their magnitude. We say, for example, an object appears to be six inches, or six feet long. Now, this cannot be meant of visible inches, etc., because a visible inch is itself no constant determinate magnitude and cannot therefore serve to mark out and determine the magnitude of any other thing. Take an inch marked upon a ruler; view it successively at the distance of half a foot, a foot and a half etc., from the eye; at each of which, at at all the intermediate distances, the inch shall have a different visible extension, i.e., there shall be more or fewer points discerned in it. Now, I ask, which of all these various extensions is that stated determinate one that is agreed on for a common measure of other magnitudes? No reason can be assigned why we should pitch on one more than another. And, except there be some invariable determinate extension fixed on to be marked by the word "inch," it is plain

and consequently affected in like sort by their senses; [of course, it does not follow that like faculties entail being affected identically; visual judgments of size vary with perspective] and who had never yet known the use of language; they would without question agree in their perceptions. Though perhaps when they came to the use of speech, some regarding the uniformness of what was perceived might call it the 'same' thing; others especially regarding the diversity of persons who perceived, might choose the denomination of 'different' thing." Unfortunately Berkeley does not distinguish here the sense of "same" when we talk of the same color and the sense of "same" when we speak of the same spatio-temporal object. Assuming that Berkeley means the latter, his point might be that various observers can agree about the apparently "same" spatio-temporal location of a datum, for example a thermometer, although realizing that in "strictness" of speech there are as many thermometers as there are observers. This is no more than to say that the metaphysics of immaterialism is irrelevant (hence no obstacle) to the practical issue of spatio-temporal individuation. Berkeley allows an "apparent" space, and we would assume could allow the "apparent" spatio-temporal individuation of a datum with respect to an agreed upon reference point in this space.

it can be used to litle purpose; and to say a thing contains this or that number of inches shall imply no more than that it is extended, without bringing any particular idea of that extension into the mind ... From all which it is manifest that the judgments we make of the magnitude of objects by sight are altogether in reference to their tangible extension. Whenever we say an object is great or small, of this or that determinate measure, I say, it must be meant of the tangible and not the visible extension which, though immediately perceived, is nevertheless little taken notice of. (*Essay* 61)

Berkeley clearly understands that determination of the metric property of length requires the ascertainment of congruence and the existence of a standard "unit" of length. Through successive applications of the "unit" we can get a measure of the extension of the boundaries of physical objects; and derivatively, a measure of areas, volumes, etc. It follows, however, that metric length is not an intrinsic property of objects (such as color, or taste or even apparent extensiveness), but rather a relational property; we are determining a relation between the magnitude to be measured and the standard unit, specifically, how many times the unit goes into the magnitude to be measured.

It is from the point of view that metric length is a relational property that Bailey criticizes what he takes to be Berkeley's claim in the above passage, that because of *apparent* size of the ruled "inch" varies with distance, apparent size is useless as a "common measure." Bailey correctly rejects this view, pointing out that it is irrelevant (except perhaps for practical purposes) what distance the comparison is made between the unit and the magnitude to be measured; if the "inch" can be laid of ten times against a magnitude to be measured at ten feet, the same will happen at twenty feet, and in either case the magnitude to be measured will be called "ten inches." [32] Moreover, he points out that the determination of congruence, essential to measurement, is normally by sight, contrary to what is perhaps a suggestion in Berkeley, that congruence (in the sense of the coincidence of end points) is ascertained tactually.

Bailey's criticisms help us focus on two problems raised in these unfortunately brief passages concerning measurement. The first concerns the nature of the standard unit of length; the second concerns the establishment of the equality of lengths. Concerning both issues, we can ask how Berkeley's remarks relate to his broad contention that metric descriptions of objects, those useful to science, deal with "tangible" magnitude.

With respect to the first problem, and ignoring for the moment ambi-

---

[32] Bailey *op. cit.*, p. 90.

guities in the concepts of "visible" and "tangible" magnitude, we can take Berkeley to be asserting that the length of our measuring rod cannot be a function of its spatial position. In more modern parlance our metrization of space presupposes that the measuring rod not change its length merely through transport; or alternatively, that if the measuring rod can be made coincident with two other segments, regardless of their spatial position, the two will be said to be equal.[33] Modern discussions make clear that the contention that the length of the measuring rod is not a function of its spatial position is, from a logical point of view, a convention; other foundations for the metrization of space are logically possible.[34]

We are taking Berkeley's phrase, "invariable determinate extension" to refer to the invariancy of the measuring rod with respect to spatial position. However, for Berkeley, the property of "rigidity" is not conventionally attributed to the rod, but somehow perceptually ascertainable through touch. Unfortunately Berkeley does not develop this point. The simplest meaning is perhaps that we generally attribute "rigidity" to objects that are "solid," and solidity is a quality revealed to touch. However, the claim that "solid" bodies are "rigid" is not itself ascertained by touch, but rather by sight, if the claim of rigidity for solid bodies is at all a perceptual claim. Within limits we have some visual awareness that a solid body does not change shape as it is transported. However, we must remember that for Berkeley apparently visual judgments of the constancy of size of moving objects are implicit judgments of their "tangible" magnitude. If this means that we are in fact judging that the use of a measuring rod will give us the same result for the magnitude of an object in two different positions, it makes no sense to make such implicit judgments of the measuring rod itself. The judgment of the constancy of the rod, is, if a perceptual judgment, one made by sight. If Berkeley's claim is, in fact, that it is objects solid to the touch which are "rigid," the claim of rigidity, if a perceptual one, is itself not made by touch.[35]

[33] See for example, Reichenbach. *The Philosophy of Space and Time,* Chapter 1, (sec. 4, 5, 8) (New York: Dover Publications, Inc., 1957), translated Maria Reichenbach and John Freund from *Philosophie der Raum-Zeit Lehre,* 1928, reprinted in A. Danto and S. Morganbesser Ed. *Philosophy of Science* (New York: Meridian Books, 1960), pp. 383-399.

[34] To say it is a convention, does not mean totally arbitrary. Such an assumption allows us to have simple laws that explain a great many phenomena.

[35] Reichenbach *op. cit.,* p. 390 discusses the distinction between the concept "solid" which can be ostensively defined and the concept "rigid" which involves as part of its meaning that the length of an object is not (ceteris paribus) a function of its spatial position.

Similar problems arise with respect to the ascertaining of congruence, of the equality of lengths. Is there any sense in which such determination of congruence is essentially related to tactual awareness? Perhaps one could claim that in the act of laying off a measuring rod against an object we are in contact (touching) the object. That is, our determination of congruence is made at *no distance* from the object. Even if this were true, Bailey's criticism is still apposite; the determination of the coincidence of end points is often (although not necessarily) made by sight. Moreover, assuming the invariance of length with respect to position, the act of measurement, or the successive laying off of the measuring rod, can be done at any distance from the observer.

In sum, then, we find no determinate sense to Berkeley's claim that the invariancy of "measured" magnitude with respect to "distance" from the observer is essentially related to tactual awareness. Although the passages quoted above concern "magnitude," our conclusions could as well apply to the concept of "tangible" distance. For if by "tangible" distance, is meant distance as measured, there are fundamentally the same issues concerning the invariancy of the "unit" through transport, and the determination of the coincidence of the end points of the rod with the segment whose length we are determining.

Berkeley constrasts the invariancy of what he calls "tangible" magnitude with the variability of "visible" magnitude; "a visible inch," as he says, "is itself no constant determinate magnitude and cannot therefore serve to mark out and determine the magnitude of any other thing." (*Essay* 61) Later in the *Essay,* and referring back to this passage, he makes the following interesting remark:

To come to a resolution in this point, (what is the object of geometry) we need only observe what has been said in secs, 59, 60, 61, where it is shown that visible extensions in themselves are little regarded and have no settled determinate greatness, and that men measure altogether by the application of tangible extension to tangible extension. All which makes it evident that visible extension and figures are not the object of geometry. (Essay 151)

We will discuss later the issue of the "object of geometry." What is of interest here is the claim that in measurement, *both* the measuring rod and that to which it is applied are said to be "tangible" extension. As we have seen, however, beyond the claim of a certain kind of invariancy (with respect to position), it is not clear to what "tangible" refers, in the phrase "tangible magnitude."

We can now raise the question whether any determinate sense can be given to the concept of "visible magnitude." Our discussion will eventually lead to an investigation of Berkeley's claim that with respect to visual perception, there exist certain irreducible visual contents termed "visible minima."

First we recall Berkeley's remark (*Essay* 61) that the "inch marked upon a ruler" will have at different distances different "visual extensions" and is therefore useless as a unit of measure. And by "different visual extensions" Berkeley understands "more or fewer points discerned in it." Berkeley's point, at least with respect to the problem of measurement, can be clarified in terms of an example. We imagine an object at a fixed distance from us, whose length is to be measured. We hold our thumb and forefinger up to one eye and (a) view the ruled inch when it is held ten feet from us. Closing our thumb and forefinger until visually they exactly enclose the "inch," we then use that distance between thumb and forefinger as a "unit," and consecutively lay it off against the object to be measured, gaining a certain measure of its length; (b) view the ruled inch when it is held twenty feet from us. The thumb-forefinger unit will become less, and therefore, the number of times it can be laid against the object will be greater, giving us a different measure of the object's length.

If our example correctly expresses Berkeley's meaning, we have no disagreement as far as measurement goes. Clearly, if different units are used, we will get different measures for the length of the object. On the other hand, if any of these thumb-forefinger units was consistently used and called the "inch" there would be no problem. Moreover there would be no problem if all the units were used, as long as the original ruled "inch" and the object to be measured are at the same distance from the eye. The variability in measurement comes when two unequal (non-coincident when placed side by side) thumb-forefinger inches are separately used to measure the length of an object which is at a *fixed* distance from the observer. For this to happen at least one of the thumb-forefinger inches must be transported in space without change, gaining the status of a "tangible" inch. The alleged dilemma of variable measurement occurs when two non-coincident "tangible" rods (invariant with respect to position) are both called the unit.

Berkeley might agree with this analysis; that all he means by the concept "visible inch" is the set of non-coincident (when placed side by side) lengths each of which is gained by enclosing the "ruled" inch between thumb and forefinger as it varies in distance from the observer. And he might well agree that *each* member of the set, if it is transcribed (say, on paper) and transported to various places to function as a measuring unit, has the status of a "tangible" unit. Support for this view is gained by looking at other places where Berkeley speaks of "visible magnitude." His reference is generally to certain geometrical or physiological considerations, and not to a metric somehow intrinsic to visual consciousness itself.

For example, Berkeley contends that

the magnitude of the visible moon, or that which is the immediate and proper object of vision, is no greater when the moon is in the horizon than when it is in the meridian. (*Essay* 74)

Although Berkeley uses the phrase "immediate and proper object of vision," his claimed equivalence for the visible extension of the horizontal and meridian moon is based on the fact that "the angle under which the diameter of the moon is seen be not observed greater in the former case than in the latter." (*Essay* 67) Yet Berkeley should admit on his own principles that strictly speaking the angle subtended by the diameter of the moon is not seen. Moreover the equivalence or non-equivalence of these angles would be established by some sort of measurement, making use of a unit invariant with respect to visual perspective; and therefore, to use Berkeley's terminology, a "tangible" unit.

It should be remarked, however, that from a definitional point of view there might be no objection to defining the "visible" extension of the moon as the *set* of angles subtended by its diameter. And if members of this set differ in degree, it is analytically true that the phrase "visible extension of the moon" does not refer to a unique number (to a "settled" magnitude). As we will see, however, it is not clear that Berkeley merely means this, rather than a metric intrinsic to visual consciousness itself.

In *Visual Language*, there is a more elaborate attempt to set up what we could term a visual metric. Berkeley conceives a "diaphanous plane," set up perpendicular to the horizon, and divided into squares. Objects in the visual field are projected onto the plane, and those objects that "occupy most squares" are said to have the greater "visible extension." Berkeley correctly points out that there is no necessary connection between this "visible extension" and our judgments of magnitude.

those which are seen through the upper squares shall appear vastly bigger than those seen through the lower squares, though occupying the same or a much greater number of those equal squares in the diaphanous plane.

(*Visual Language* 55)

Yet Berkeley, apparently recognizing the difficulties in his concept of "visible extension" notes that the "diaphanous plane and its images are

altogether of a tangible nature." (*Visual Language* 57). However, after
admitting this, he goes on to claim that the "visual objects" (which he
calls "pictures") have an order analogous to the "tangible" images on
the screen, with respect, for example, to the designations "higher" and
"lower." There is, however, no such analogy. An object visually considered to be more distant than another will project higher on the diaphanous plane, but the objects do not appear visually in that order.
This only happens after the projection is made.[36]

What Berkeley has done, in fact, is given a tangible analogue to what
he considers to be the *uninterpreted* visual field, that primitive visual
experience (allegedly experienced by the "born blind" who regain sight)
in which all objects are at "no" distance from the eye. Assuming some
cognizance of spatial relations, all objects in the field would in truth
appear only in the relations left-right or upper-lower; exactly the relations
of subjects on the diaphanous plane, or for that matter, on the retinal
image. Although it would be wrong, as we have previously said, to charge
Berkeley with explicitly identifying the structure of this primitive visual
experience with the planar structure of the retinal image,[37] we are always
faced with the problem that as Abbott and Bailey suggest, he offers no
compelling evidence that "tri-dimensionality" is not a characteristic of
all visual experience.

If we wish, then, to make intelligible the notion of a metric intrinsic
to vision, we should reject such "measures" of visible magnitude as the
angle subtended by the object (with the eye), or the number of squares
occupied on the diaphanous plane, or the proportion that the "image"
of the object takes up on the retina. Berkeley does, we believe, offer some
suggestions in the attempt to make sense of such a metric, and to these
we now turn.

In the *Essay*, we recall, Berkeley speaks of the ruled inch having a
different visible extension as its distance from the observer varies, or, as
he says, "there shall be more or fewer points discerned in it." (*Essay* 61)

---

[36] Abbott (*op. cit.*) makes the point that with respect to objects at relative distances
from us, we can by moving our head in the same plane up or down or right or left,
alter the relative up, down and right, left relations of objects at different distances
from us. We cannot do this with objects that are in the same plane perpendicular to
our line of sight. Abbott believes such parallax phenomena are evidence for the
existence of three dimensional space. As far as we know parallax phenomena are not
dealt with by Berkeley.

[37] The closest he seems to come to such an identification is the early *Philosophic
Commentaries*: For example, No. 169: "The sphere of vision is equal whether I look
only in my hand or on the open firmament for 1st, in both cases the retina is full;
2nd, the radius of both spheres are equal or rather nothing at all to ye sight; 3rd,
by equal number of points in one and t'other."

This suggests that the visual field is composed of minima ("points"), and we can compare two visual objects with respect to the number of points they contain. Moreover, as we will see, the suggestion is that these minima are not ascertained by a process extrinsic to visual consciousness itself (determining the number of squares occupied on the diaphanous plane would be such an "extrinsic process"), but rather their recognition is constitutive of such consciousness.

We should remark that the doctrine that there exist sensible minima is an important component of Berkeley's empiricism, having important ramifications in his philosophy of mathematics, particularly his views concerning infinity. We will later deal with these. With respect to the concept of visible minima, Berkeley's views are elaborated in sections 80 to 86 of the *Essay*, *Principles* 124, and various entries in the *Philosophic Commentaries*. These can be summarized as follows:

a. No sense content (or object of sense, or "idea") is infinitely divisible; with respect to vision, therefore, there must be a minimum component of the visual field.

b. Any finite extension (visible or tangible) must be composed of a finite number of minima.

c. The "minimum visible" is the same for all beings with the "visive faculty."

It is clear that (b) is merely a logical consequence of (a) if in fact it is meaningful to assign a magnitude to a visible minimum. It the size of the parts of a finite aggregate has a lower limit possessed by at least one part, the aggregate consists of a finite number of parts. It is not clear, however, in what sense we are to assign a magnitude to each minimum, since there is no intuitive anwareness of the visual field as an aggregate of minima; rather the field has the apparent quality of continuity, not discreteness.[38] Nor can the minimum have the character of what was called in the development of the calculus an "infinitesimal," that is a magnitude greater than zero but less than any finite quantity; since (aside from other reasons Berkeley will offer against their existence) minima of "infinitesimal" magnitude would allow finite extensions to be composed of an infinite number of minima. However, (c) unfortunately uses metric language to express the view that different visual consciousness in order to be compared requires a procedure of measurement extrinsic to each visual consciousness. For example, the claim that my visual threshold (before the object disappears from view) is a width

[38] See *Philosophical Commentaries*, No. 321: "Question: why so difficult to imagine a minimum. Answer: because we are not used to take notice of em singly, they not being able singly to pleasure or hurt us, thereby to deserve our regard."

of one tenth of an inch, whereas yours is one-twelfth, presupposes a metric for the object invariant with respect to both our visual perspectives. Calling the minima of two visual consciousness the "same" in magnitude, however, suggests they can be compared. To suggest the same thought more in the language of immaterialism; the "minimum" for me does not refer to some object that exists independent of vision, and whose magnitude could be determined in some other way (with a ruler for example), therefore it is a meaningless (not false) to say it has parts.[39]

It is to (a) however that we will direct most of our attention. Is it meaningful, and if meaningful is it true to say that a visual content cannot be infinitely subdivided? One point that should be noted in the beginning. To base the existence of minima on the impossibility of infinite subdivision, is quite different from basing it on some claim that we intuitively recognize visual contents as an aggregate of points. And Berkeley's stronger arguments for the existence of minima are those that are based on the former, rather than the latter consideration.

Some of Berkeley's commentators evidently feel there is no problem in the claim that a perceptual content cannot be infinitely subdivided. For example, Wild simply states: "A conscious experience cannot be infinitely subdivided." [40] Unfortunately Wild does not in any precise way say what is meant by the operation of subdividing a perceptual (conscious) content, nor on what grounds we would claim there cannot be an infinite subdivision of such a content. Armstrong comments on the same point.

Since there can be nothing hidden in our sense impressions, and since the capacities of our senses is not infinite, these sense impressions cannot be divisible without limit.[41]

[39] In the *Essay* (81) Berkeley, rejecting that the "minimum visible" of a "mite" can be less then a "man," says: "To which I answer, the minimum visible having (in like manner as well other proper and immediate objects of sight) been shown not to have any existence without the mind of him who sees it, it follows there cannot be any part of it that is not actually perceived and therefore visible." The quality, then, of being a minimum is a relational property with respect to visual consciousness as a whole. To compare the "magnitude" of sections of two different consciousnesses presupposes some characteristic shared by both sections like length or area, which can be compared. And the comparison would have to be made by a method (measurement of corresponding retinal images, for example) which is extrinsic to the visual consciousness. An analogy might be drawn with the experience of pain. If intensity of pain merely refers to a certain quality of experience itself, this quality which is intrinsically private cannot be revealed to another for comparison. It would be mistaken, however, to conclude from this that the minimum pain intensity is the same for all. If it illegitimate to use comparative terms like "greater" or "smaller," it is equally illegitimate to use the term "equal." The same is true for visual minima; it is misleading to say that the minimum for one consciousness is equal to the minimum for another, since comparison is in principle impossible.

[40] Wild, *op. cit.*, p. 98.

[41] Armstrong, *op. cit.*, p. 44.

Now there is a clear sense in which we can talk about the limitation of our perceptual capacities. There may be finite extensions that we cannot see unaided, but can see with a microscope. We may not be able to "see" anything smaller then a certain number of inches, or hear sounds higher than a certain number of frequencies. A presupposition for this kind of analysis, however, is that we consider it proper to speak of qualities of objects whose degrees can be ascertained in ways other than the particular perceptual consciousness which allegedly recognized the minimum. We might call these threshold minima; by which we mean a certain degree of a quality, such that when the degree of the quality is less than this, the quality cannot be recognized by a particular sense. There is no difficulty about using metric locutions about such minima; we can say, for example, that when an object at distance (d) has its area reduced less than (x) square inches, it cannot be seen by the unaided eye. The minimum area with respect to the unaided eye can then be called (x) square inches.

Threshold minima, however, are not what is meant by either Armstrong or Berkeley. For Berkeley such minima would be relative to the visual acuity of individual perceivers, and hence violate condition (c) above. Moreover the concept of threshold minima presupposes that there be a method of establishing the degree of a certain quality in an object other than the perceptual sense for which there is the claimed threshold. It is because in principle I can determine by measurement that an object is less then (x) square inches in area, that I can say that (x) square inches is the visual minimum or threshold. The implication is that the minimum visible has parts which (since they are less then (x) square inches) are not seen, and this possibility is denied by Berkeley. It is denied because, for Berkeley, the minimum is established within visual consciousness itself, not through some metric extrinsic to vision; minimality, one might say, is itself a quality of part of the field and recognized as such by vision.

Berkeley developed his argument for the existence of sensible minima in some detail, in the *Principles*.

Every particular finite extension which may possibly be the object of our thought is an idea existing only in the mind, and consequently each part thereof must be perceived. If, therefore, I cannot perceive innumerable parts in any finite extension that I consider, it is certain they are not contained in it; but it is evident that I cannot distinguish innumerable parts in any particular line, surface, or solid, which I either perceive by sense, or figure to myself in my mind: wherefore I conclude they are not contained in it. Nothing can be plainer to me that the extensions I have in view are no other than my own ideas; and it is no less plain that I cannot resolve any one of my ideas into an infinite number of other ideas, that is, they are not infinitely divisible. (*Principles* 124)

The argument that because I cannot *distinguish* an infinite number of parts in a finite extension (say, a bounded patch of color) it must contain a finite number of parts, seems clearly mistaken, even accepting the premise that every part (qua "idea") must be perceived. I cannot distinguish an infinite number of parts in a patch of color, because it appears as a continuous manifold and not as an aggregate of parts, finite or infinite in number. Some manifolds can appear as discrete, that is as aggregates, and one might ask about these whether an infinite number of parts can be apprehended in a single consciousness? Other manifolds appear continuous, that is as not containing parts.

Berkeley's argument, however, might be strengthened, if we could give a precise meaning to the notion of resolving or dividing a sense content, and somehow demonstrate that such a resolution or division cannot be carried on and infinitum. We find, interestingly, in Hume an example of such a process of dividing a sense content, which is used to make an identical point, that there cannot be an infinite division of the objects of sense. It should be pointed out that by object of sense or "impression" Hume does not mean the object that produces the "impression" (Hume unlike Berkeley might allow the existence of such objects) but like Berkeley, sense object means object as content of consciousness. Hume writes in the *Treatise.*

Put a spot of ink upon paper, fix your eye upon that spot, and retire to such a distance that at least you lose sight of it; it is plain, that the moment before it vanished, the image or impression, was perfectly indivisible.[42]

We will assume Berkeley would have accepted such an example, and therefore direct our criticism to the issue whether this example or comparable ones can establish with any finality that sense contents cannot be divided ad infinitum.

To delineate the issue quickly, we will present an alternative thesis to Hume's: that as the distance from the "spot" increases, there is a continual diminution of the "impression" to zero, that is, until it disappears. By a continuous diminution to zero we mean that for every apparent

[42] David Hume, *Treatise,* Vol. 1, London; Dent (Everyman's Library), 1911 (originally published in 1738), p. 35.

Hume confuses the issue by first considering the impossibility of infinitely dividing the objects of imaginaton. The mages, he claims, of a thousandth and a ten thousandth part of a grain of sand are not different. Neither we might add are the images of a four and a five foot extension. I can have images of the equality and inequality of lengths, but it is inappropriate to use metric locutions about images.

magnitude between the original impression of the spot and no impression, there exists a lesser apparent magnitude.

The question becomes, on what grounds can we decide between the two views? Hume appears to believe it is an empirical question, the evidence coming from introspection; we are aware that there is an apparent magnitude (M) such that no apparent magnitude would be judged less then (M). Certainly there are sequences that have the characteristic of what we can call discrete diminution to zero; at one moment the impression or sense datum seems to have a certain extension, and immediately thereafter it seems to have disappeared. However, the introspective evidence does not seem to rule out a continuous diminution to zero. Just as we are aware of a continuous motion (with no apparent jumps) we can be aware of a continuous diminution of a visual datum to zero. The two can often go together when, for example, a moving object gradually covers up a patch of color until the latter is completely hidden.

The continuous diminution of a visual datum to zero is consistent, it would seem, with both of Armstrong's reasons for the existence of visual minima; (A) that "there is nothing hidden in our sense impressions; and (B) "the capacity of our senses is not infinite." In our judgment (B) properly applies to the existence of threshold minima, and has no obvious relevance to the issue of whether a visual sense datum ("impression") can be infinitely reduced.

If the psychological evidence is not conclusive, could logical grounds be offered for the existence of visual minima? One could claim that if there was no minimum visible, the visual datum could not disappear from consciousness, since an infinite number of apparent magnitudes (each less than the other) would have to be passed through. Conversely, if the visual datum disappears from consciousness, the diminution must occur in a finite number of steps, the datum immediately preceding zero being the minimum. Presented in this way, the argument is an example of the Zenonian paradox with respect to an allegedly continuous and finite motion; that such a motion requires the achievement of an infinite number of states in a finite time (each state succeeding the other).

We have no evidence that Berkeley used such an argument against the claim of a continuous perceptual diminution. As with Hume, Berkeley's argument (in the *Principles*) for the existence of a visible minimum is supported by the alleged data of introspection. We do, however, have a good deal of evidence, that Berkeley was concerned with what may be called paradoxes of the continuum, some of which are examples of the Zenonian paradoxes of motion, and we know he felt these para-

doxes followed from the claim that "finite extension" was infinitely
divisible. "It is," he claims in the *Principles,*

the doctrine of the infinite divisibility of finite extension . . . (which) is the
source from whence do spring all those amusing geometrical paradoxes which
have such a direct repugnancy to the plain common sense of mankind, and
are admitted which so much reluctance into a mind not yet debauched by
learning . . . Hence, if we can make it appear that no finite extension contains
innumerable parts, or is infinitely divisible, it follows that we shall at once
clear the science of geometry from a great number of difficulties and
contradictions which have ever been esteemed a reproach to human reason.
(*Principles* 123)

Berkeley, as far as we know, nowhere gives an explicit and detailed
treatment of these "amusing geometrical paradoxes," but we would list
three alleged paradoxes either explicitly mentioned by him or implicit
in his detailed critique of the calculus. The first is the paradox of the
equivalence of part and whole in an infinite series.[43] One might object
that this has nothing uniquely to do with infinite divisibility, but a
property of two denumerably infinite sets (for example, the natural
numbers and the even natural numbers) that they can be put into one-
to-one correspondence. More important are two alleged paradoxes concerning the conception of extensive magnitudes as "continuous." We
are using "continuous" somewhat loosely here to refer to what would
now be called the property of "denseness"; that for any two distinct values
of the property in question, there is one value greater than the least of
the two and less than the greater. Both "paradoxes" concern what is
called the "actual infinite," the assumption that finite extensive magnitudes are not only potentially divisible *ad infinitum,* but actually composed of an infinite number of parts.

The first "paradox" is the non-kinematic paradox of composition,
and it is illustrated in the concept of the "infinitesimal." If a finite extension is considered composed of an infinite number of parts, each
part itself could have no determinate magnitude; hence an "infinitesimal." The paradox refers to the problem of how any determinate
magnitude could be composed of parts which themselves have no determinate magnitude. In the *Analyst,* and related writings, Berkeley
subjects the concept of the infinitesimal to logical criticism, although the
problem of the paradox of composition is not stressed.[44]

---

[43] In the *Philosophic Commentaries,* there is the following remark: "Mem to prove
against Keill that the infinite divisibility of matter makes the half have an equal
number of equal parts with the whole." (J. Keill (1671-1721) Author of *Introduction
ad Veram Physicam*)

[44] Berkeley rejects as unintelligible the concept of the infinitesimal, and therefore

The second "paradox" can be expressed kinematically; how can a moving object achieve an infinite number of states in a finite time? An alternative and geometrical formulation concerns the problem of considering a circle as a regular polygon whose sides have been doubled an infinite number of times. A more general and mathematical formulation, and one central to the controversy engendered by Newton's work in Analysis, is whether the limit of an infinite series can also be a member of the series.

There is, as we have mentioned, no detailed and explicit discussion of these "paradoxes" in Berkeley's writings. They do form, however, part of the historical ground or impetus for Berkeley's sensible arithmetization, or better, sensible finitization of finite extensive magnitudes; at the heart of which doctrine is the conception of sensible minima. There are three related doctrines at the core of this program of finitization:

a. Extensive magnitudes are perceptual magnitudes, either visual or tangible. For Berkeley this means no extensive magnitude can exist independent of perception. "Space" itself, or "extension" itself is a meaningless term referring to nothing in reality; it is vacuous, though not self-contradictory.

b. Finite sensible extension (visible or tangible) is composed of finite aggregates of sensible minima.

c. Geometry as a "science of magnitude" deals with (or has as its "object") perceptual extensive magnitude (the *Essay* suggests it is "tangible" magnitude).

There are certainly serious problems in the program, particularly related to Berkeley's conception of geometry. For example, (c) suggests something like the Aristotelian view that geometry as a science deals with properties of the boundaries of physical objects, properties which are conceptually, though not physically separable.[45] We must remember, however, that for Berkeley such "physical" boundaries are not continuous, but aggregates of minima. We will deal later with this philosophy of

---

the minimum magnitude of any extensive segment would be assumed to have some determinate magnitude greater than zero. It is an elementary consequence that the number of such minima for any finite total extension would be finite. One might offer a third possibility, that the finite segment, for example a unit interval, is composed of an infinite number of parts, each of which is a determinate magnitude greater than zero, in this case the infinite series $\frac{1}{2} + \frac{1}{4} + \frac{1}{8} \ldots + \frac{1}{2}n$ (n = 1, 2, 3, ...). The intuitive problem here is somehow conceiving an actually bounded length, which is said to have no last part. Kinematically the problem is that of the "Achilles," where it appears difficult to comprehend how an actually infinite number of motions can be accomplished in a finite time.

[45] There is a suggestive remark along these lines in the *Philosophic Commentaries*, (119): "Lines and points conceived as terminations different ideas from those conceived absolutely."

geometry, now merely pointing out that, however successful his program, he raises the fundamental issue concerning the application of mathematics to the study of physical objects. In capsule form the issue can be expressed as follows: How does one determine whether the integers, the rationals, or the reals should be held isomorphic to physical magnitudes? Alternatively one could ask, how do we determine whether physical magnitudes have the property of discreteness, denseness, or continuousness? Berkeley appears to opt for "discreteness," which raises serious questions of how one applies Euclidean geometry to such manifolds.

It would be helpful to summarize our remarks concerning the thesis that there exist sensible minima, focusing on visible minima, although many of the comments could be applied to the alleged minima of other senses.

By a "visible minimum" Berkeley does not mean what we have termed a "threshold minimum," that is a minimum for vision of a certain magnitude, where the measure of this magnitude is not established by vision. When I say that at ten feet the minimum length I can visually perceive is one eighth of an inch; that fact (the measure of the length) is not established by vision. Although remarks in the *Philosophic Commentaries* and elsewhere often suggest that Berkeley identifies the "minimum visible" with "points" on the retinal image,[46] his more mature view is that the quality of "minimumness" or indivisibility with respect to a datum of vision is grasped within visual consciousness itself. As Wild suggests, the concept of a visible minimum is a relational concept; certain items are apprehended as a minimum in relation to the total visual field. We might consider the retinal image as an anlogue to the visual field, "points" on the retina correspond to minima in the field.[47]

We observed, however, that the evidence from introspection for the existence of such minima seems inconclusive. Hume's experiment concerning the diminshing spot (and analogous experiments) is not, in our judgment conclusive. Particularly we question whether the data from introspection rule out an apprehended continual diminution of a visual "patch" to zero (disappearance). We understand continuity here as one understands the property of "denseness" with respect to the rational number; for every visually apprehended magnitude of the "spot" there

---

[46] *Philosophic Commentaries* (169) "The sphere of vision is equal whether I look only in my hand, or in the open firmament, for in both cases the retina is full."

[47] Wild, *op. cit.*, pp. 88-89. Berkeley himself remarks in the *Philosophic Commentaries* (204) "The greatness per se perceivable of the sight, is only the proportion any visible appearance bears to the others seen at the same time; or (which is the same thing) the proportion of any particular part of the visual orb to the whole."

exists one that is apprehended as less. We acknowledge that our claim of *no minimum* is embedded in the very meaning of the concept of perceptual continuity; that is we cannot in fact reconstruct the apprehension of the infinite number of magnitudes as a step by step process. On the other hand the claim for the existence of "minima" is embedded in the apprehension of perceptually discrete sequences. And sense phenomena as often appear continuous as discrete. We have noted that Berkeley's program to finitize bounded segments of physical extension, is impelled in part by the "paradoxes" connected with conceiving finite wholes as composed of an actually infinite number of parts; either a finite extension composed of an infinite number of pieces, or a finite motion composed of an infinite number of distances traversed. Berkeley's critique here is historically rooted in the problem of understanding the concepts of *continuity* and *limit* independently of geometrical or kinematic intuition. The concept of the "infinitesimal" found incoherent by Berkeley; whether conceived statically as an indivisible segment of length, or dynamically as an "impetus" or kinematically as an "instantaneous velocity"; is one attempt to solve this problem. Berkeley's own arithmetization of the continuum based on the notion of sensible minima can be viewed in part as a substitute for the arithmetization based on the concept of the "infinitesimal."

Assuming the existence of visible (or tangible) minima, there is still the problem of using such minima for measurement: making them the object of a metric geometry. For example, it can be asked whether the individual minima have magnitude, since, though indivisible, they are not "infinitesimals." In the *Philosophic Commentaries* (No. 273), Berkeley says:

whether m. v. (minimum visible) or T (tangible minimum) be extended.

Berkeley never resolves the question, perhaps recognizing that it cannot be resolved. Since the minimum is the fundamental measure of magnitude, with respect to a particular sense, there is no way within that same faculty of sense of determining the magnitude of an individual minimum. If extended, it is in an irreducibly intuitive sense, and it is meaningless to attach some number to its magnitude. Moreover, since we do not apprehend either visible or tangible magnitudes as aggregates of minima, it is difficult to see how they could be used as the foundation of an intrinsic metric of finite segments of extension, in terms of the number of minima they contain. With respect to the field of vision, we can, of course, make tentative judgments about the proportion a given visual datum occupies in the visual field. ("One half," "One fourth," etc.) This is quite different from a visual metric in terms of the number of "minima" contained by a datum.

With respect to such "proportional" judgments, we have no quarrel with Berkeley's claim that there is no necessary connection between the relative proportions of the visual field occupied by two items and their relative sizes. As regards the latter, there is an invariancy with respect to visual perspective, an invariancy at the core of the concept of the "primary quality" of length. However, as we have mentioned, Berkeley's attempt to define this invariant metric in terms of the data of touch is ultimately unsatisfactory. One could perhaps suggest that what Berkeley means is that when I touch the boundary of an object, I am at "no distance" from it and therefore the number of "tangible points" apprehended is clearly invariant with respect to visual perspective. Yet we do not apprehend the boundaries of physical objects necessarily as aggregates of "points" or minima; thus this suggestion would appear useless as far as a theory of measurement is concerned.

Somewhat comparably, since we do not apprehend portions of the visual field as aggregates of points, it would seem fundamentally mistaken to speak of visual "size" or "magnitude," if we mean by that, attaching numbers to portions of the visual field corresponding to the number of "points" or minima they contain.

When we turn to Berkeley's Philosophy of Mathematics, particularly his critique of the calculus, we will again be concerned with the role of this concept of "sensible minima." Preceding that, we will investigate some important aspects of his Philosophy of Physics, beginning with his analysis of the concept of "matter."

THE PHILOSOPHY OF PHYSICS

A. THE CONCEPT OF MATERIAL SUBSTANCE (MATTER)

Berkeley, as we have mentioned, finds the expressions "material substance" to be meaningless. However, a reading of the texts in which the concept of "material substance" is closely analyzed, reveals three distinct, although related, issues that are involved in this claim.[1] The first (1) we will label the issue of "property inherence," or how material substance is related to its predicates; the second (2) we will call the issue of "mind-body interaction," or how "material substance" can be understood to cause or produce our mental states. (2) can also be construed as the issue of how the "insensible" (or non-observable) particles postulated in atomic theories can cause or produce the observable macroscopic phenomena we are acquainted with through our senses. (2) construed in the second way allows the possibility that the phenomena revealed to our senses are often actual qualities of independently existing macroscopic objects (for example, color), but raises the question of how the interaction of the "insensible" particles can produce such "qualities." The third (3) issue, phrased somewhat simply and crudely is, how can "material substance" (construed as "insensible" particles) exist; where the question is not understood as how can something exist apart from its properties, but rather, how can any non-spiritual and unperceivable object exist.

Issue (1) is not uniquely related to questions concerning the existence of the particles postulated in atomic theories. That is, the issue of property inherence is as much an issue (if it is one at all) with respect to how the electron is related to its mass, as it is with respect to how the table is related to its color. Yet in the *Principles* and the *Three Dialogues,* it is

---

[1] The basic texts referred to here are the *Principles of Human Knowledge,* and *The Three Dialogues Between Hylas and Philonous.*

this question of property inherence which is often the focus of the discussion. In the *Three Dialogues,* for example, we have the following exchange:

Philonous : "Material Substratum" call you it? Prey, by which of your senses are you acquainted with that being?

Hylas : It is not itself sensible; its modes and qualities only being perceived by the senses.

Philonous : I presume then it is by reflection and reason you obtained the idea of it?

Hylas : I do not pretend to any proper positive idea of it. However, I conclude it exists because qualities cannot be conceived to exist without a support. (*Dial* 1-137)

Pressed to explain in what sense "material substratum" supports its "modes" (like "extension"), and admitting that it is not in the literal sense that "your legs support your body," Hylas is reduced to concluding that he does not understand "what was meant by matter's supporting accidents." The discussion in the *Principles* (16) with respect to the vacuousness of the term "support" is comparable, after which Berkeley remarks:

If we inquire into what the most accurate philosophers declare themselves to mean by "material substance" we shall find them acknowledge they have no other meaning annexed to those sounds but the idea of being in general together with the relative notion of its supporting accidents. The general idea of being appears to me the most abstract and incomprehensible of all other; and as for its supporting accidents, this, as we have just now observed, cannot be understood in the common sense of those words; it must therefore be taken in some other sense, but what that is they do not explain. So that when I consider the two parts or branches which make the signification of the words "material substance," I am convinced there is no distinct meaning annexed to them. (*Principles* 17) [2]

Focusing only on the concept of "material substance" as "substratum," or supporter of "accidents," then, from Berkeley's point of view there can be no "idea" or sensible mark of substance; not, however, because

---

[2] The "accurate philosopher" referred to is John Locke, the *Essay Concerning Human Understanding*; Book 2 Ch. 23, "Of Our Complex Ideas of Substance." In our judgment the concept of substance for Locke is more a logical category than an object of sense. Unfortunately his own discussion suggests that "substance" refers to something almost physically distinct from its accidents, allowing Philonous with some irony to speak of "pillars supporting temples"; or for Berkeley to speak in the *Principles* of "legs supporting bodies" as examples of analogies to what is meant by substance "supporting" its accidents. Locke's expression is as follows: "The idea, then, we have, to which we give the general name 'substance' being nothing but the supposed, but unknown support of those qualities we find existing, which we imagine cannot subsist sine re substantia; which according to the true import of the word, is in plain English, "standing under," or "upholding."

the concept is self-contradictory, but because it is vacuous.[3] The error of abstraction here is not positing a single entity as referent for an expression, when, in fact, there is only divided reference; but rather positing an entity as referent for a term when *no* such entity (even in the sense of divided reference) could exist. There are no "instances" of "material substance," much less "material substance" (or "being") in general.

In a recent and illuminating article, Jonathan Bennett agrees in the main with Berkeley's critique of Locke's discussion of "substance" as substratum, and locates Locke's problem in his illegitimately converting a question about the logic of subject-predicate statements into a question about a possible empirical referent for "substance" simpliciter.[4] (Substance as substratum or supporter of qualities.) Bennett's argument, somewhat simplified, is as follows:

To say that there is some special property that substances have, is to designate a property (S = substantiality).
Since all properties must be instantiated in a substance, (S) must be instantiated in some substance (S').
To continue in this way leads to an obvious infinite regress, suggesting that the original demand for some empirical referent for "substance" simpliciter is misguided.[5]

Bennett summarizes his view as follows:

There are many kinds of things, but things do not form a kind. There is,

[3] From another perspective, Berkeley will claim that the expression, "material substance" is self contradictory, since it joins two logically incompatible characteristics: (a) it is said to have an existence independent of mind (it is an "unperceiving" entity); and (b) it is the supporter of "accidents" or qualities such as extension, figure and motion, which are, like the secondary qualities, mind-dependent. "Hence" as Berkeley says, (*Principles* 9) "it is plain that the very notion of what is called "matter" or 'corporeal substance' involve a contradiction in it." Assuming that immaterialism is false, Berkeley could claim (and indeed suggests) that the concept of "substance" is unintelligible and hence vacuous in reference. As critics like Bennett point out, this logical critique of the concept of substance in never fully isolated from the more specialized critique of the concept of "material substance."

[4] Jonathan Bennett, "Substance, Reality and Primary Qualities," *American Philosophical Quarterly*, (Vol. 2, 1965) Reprinted in C. B. Martin and D. M. Armstrong ed., *Locke and Berkeley, A Collection of Critical Essays*, (Garden City, New York: Anchor Books) (Doubleday and Co.; 1968) pp. 86-125. Bennett admits (p. 88) Locke is not clear whether "substance" is merely a grammatical category, a short hand of the 'x' in the locutions there exists an 'x' with such and such properties, or whether it is a "psychological" category, something the mind requires to make intelligible the unity of the qualities in the thing.

[5] Bennett *op. cit.*, p. 89. Bennett refers to Leibniz (*New Essays* 11, xxiii, 2) as a source of his argument, and he quotes the following interesting passage. "In distinguishing two things in (any) substance, the attributes or predicates, and the common subject of these predicates, it is no wonder that we can perceive nothing particular in this subject. It must be so, indeed since we have already separated from it all the attributes in which we could conceive any detail."

perhaps, a concept of subject in general; but it is to be elucidated in terms of the way in which more special concepts function in certain kinds of statements, and is not to be regarded as a concept which picks out a class of items.[6]

Berkeley would accept this view. For example, the well known passage in the *Principles*:

A die is hard, extended and square (and) they will have it that the word "die" denotes a subject or substance distinct from the hardness, extension, and figure which are predicated of it, and in which they exist. This I cannot comprehend; to me a die seems to be nothing distinct from those things which are termed its modes or accidents. And to say a die is hard, extended and square is not to attribute those qualities to a subject distinct and supporting them, but only an explication of the meaning of the word "die." (*Principles* 49) [7]

Moreover, Berkeley uses a similar if less general argument concerning the question of "infinite regress" (overlooked by Bennett) if one seeks some empirical referent for "substance" distinct from its qualities. The argument is found in the following passage from the *Three Dialogues*:

Philonous : Well then, let us examine the relation implied in the term
            "substance." Is it not that it stands under accidents?
   Hylas : The very same.
Philonous : But that one thing may stand under or support another, must
            it not be extended.
   Hylas : It must.
Philonous : Is not therefore this supposition liable to the same absurdity
            as the former? (*Dial.* 1-139)

The "absurdity" referred to, as Berkeley (through Philonous) expressed it was:

every corporeal substance being the substratum of extension must have in itself another extension by which it is qualified to be a substratum, and so on to infinity? (*Dial.* 1-138) [8]

---

[6] We could offer another example of this type from language. There are many kinds of terms in a language, nouns, verbs, adjectives, etc., but "term" is not a kind of term.

[7] The argument here is insufficient, for it refers only to the defining characteristics of the "die." If we call a die "white," we are not even in part explicating the meaning of the term "die."

[8] Locke himself suggests the problem of the infinite regress. "If anyone should be asked, "What is the subject wherein color or weight inheres?" he would have nothing to say but, "The solid extended parts." And if he were demanded "What is that solidity and extension inhere in?" he would not be in a much better case that the Indian before mentioned, who saying that the world was supported by a great elephant, was asked, "what the elephant rested on" to which his answer was, 'a great tortoise': but being again pressed to know what gave support to the broad backed tortoise, replied – "something, he knew not what."

Although the particular property "extension" is referred to, it is, in fact, an example of the general argument that if substance devoid of properties is itself characterized by a particular quality (S), then a new substance (S') is required in which (S) is instantiated.

Bennett goes on, however, to accuse Berkeley of "conflating" what we have called issue (1) (the question of property instantiation), with one formulation of what we have called issue (2) (how the "insensible particles" posited in corpuscularian theories are causally related to our "ideas" or sensations).[9] Bennett argues that the fact that Berkeley focuses his attack on the concept of "*material substance*," rather than, as Locke, on the "idea" of "*substance*," is evidence that the two quite distinct issues (1) and (2) have indeed been "conflated." From a textual point of view, Berkeley's addition of the modifier "material" to "substance" does not seem as significant to us as it does to Bennett. Locke's *general* discussion of the concept of substance is in terms of an idea we have as he says, of "the support of such qualities which are *capable of producing simple ideas in us*." (*Essay Concerning Human Understanding* 23-2) Locke is clearly discussing the alleged "support" of the "primary qualities," (extension, figure, motion, etc.), those qualities of objects causally related to the occurrence of sensation. The modifier merely serves to distinguish "material" from "spiritual" substance and does not represent a "conflation of two senses of the term "matter," (as substratum, and as insensible particle). In addition, the passages in Berkeley referred to by Bennett to substantiate his claim give little evidence for it, at most indicating some vagueness about which issue the former is referring to.[10]

Bennett's essay does, however, elucidate a characteristic shared by

<hr />

[9] Bennett, *op. cit.*, p. 92.

[10] We discuss section 17 in our text (p. 87). Other passages where Bennett alleges the "conflation" takes place are 16, 37, 74, and 76. With respect to section 16, there appears to be no conflation; Berkeley does refer to "substance" as matter, but his argument concerns the intelligibility of the concept of "substratum" or "support." Sec. 37 as far as we can judge appears also to be directed against the notion of substance as substratum. "If the word 'substance' be taken in the vulgar sense, for a combination of sensible qualities, such as extension, solidity, weight and the like; this we cannot be accused of taking away. But if it be taken in a philosophic sense, for the support of accidents or qualities without the mind; then I acknowledge that we can take it away, if one may be said to take away that which never had any existence, not even in the imagination." Admittedly there is some ambiguity here; the passage could be referring to the unintelligibility of the concept of substratum as "support," or the mistaken view that properties which are mind dependent can inhere in a non-sentient substance. We think the former sense is meant, since Berkeley speaks of the "philosophic sense" of "substance." It is not clear to us how Bennett is interpreting section 74. With respect to section 76 Berkeley here is referring to the other sufficient ground for rejecting the existence of "material substance," that it would have contradictory properties; it would be said to lack sentience, and yet its "accidents" would be "ideas" or objects of sense.

"matter" as substratum and "matter" as the insensible particles of corpuscular theories, a characteristic that might lead one to conflate these two senses of the term. Both meanings refer us to something behind the scenes, something unobservable and related in some fundamental sense to the observable qualities of objects. And it is this shared characteristic which might account for some ambiguity in interpreting a particular Berkeleian critique of the concept of "material substance." Bennett is of course correct, however, in insisting that the type of "unknowability" that relates to "matter" as "substratum" (Locke's "(I) know not what") must be distinguished from the "unknowability" that relates to "matter" as the atomic structure of physical objects (Locke's "real essences," or the "internal constitution or *unknown* essence" of matter.) [11]

As we have mentioned, problems in how we are to understand the relation of "substance" to its "accidents" do not uniquely concern "insensible particles." What underlies and supports the shape of a photon is no more nor less of a problem than what underlies the yellowness of the table. It is interesting to note that Berkeley's discussion of "substance" as substratum, although philosophically important, is presented somewhat as an afterthought in both the *Principles* and the *Three Dialogues*. In both works the fundamental discussion of "substance" as substratum takes place after the discussion of "secondary" and "primary" qualities.[12] Berkeley's point in the latter discussion is that the "primary qualities" ("solidity, extension, figure and mobility") [13] are no less mind dependent ("ideas") than the "secondary qualities." [14] The import of this conclusion is that there cannot be any "unperceiving" substratum in which "extension," "figure," etc., inhere, and therefore Locke's concept of "substance" as that which underlies and "supports" extension is self-contradictory. More precisely the contradiction lies in conjoining the properties of (a) being the support of extension, solidity and the like, and (b) being non-mental. "For an idea to exist in an unperceiving thing" is, says Berkeley, "a manifest contradiction." (*Principles* 7) [15] Stripped of its properties, "material substance" becomes a mere *flatus vocis*; even Hylas would have to admit that if "matter" no longer supports or underlies the primary qualities of extension, figure, and motion, it no longer serves any useful purpose.

[11] Locke, *Essay*, Book 2 Chapter VIII, sec. 9.

[12] The criticism of material substance in terms of the intelligibility of the concept of "support" begins in *Principles* Sec. 16, 17. In the *Three Dialogues*, the comparable criticism is found in Dialogue 1, p. 138-139, Turbayne ed. *op. cit.*

[13] *Principles* 10, 11. *Three Dialogues* (Dialogue 1, p. 127, Turbayne ed.)

[14] *Ibid.*

[15] See *Principles*, sec. 73, 74.

The demonstration by Berkeley that the alleged relation between material substance and its modes, that is the relation of "support," is incomprehensible, is merely an independently sufficient reason for finding the concept "material substance" devoid of meaning. It is certainly not, in his view, a necessary condition:

But why should we trouble ourselves any further in discussing this material substratum or support of figure and motion and other sensible qualities? Does it not suppose *they* have an existence without the mind? And is this not a direct repugnancy and altogether inconceivable? (*Principles* 17)

Bennett, however, finds in the above passage a clear example of Berkeley's "conflation" of a critique of substance as substratum with a critique of Locke's realism. (The beginning of sec. 17 is a critique of the concept of substance as substratum). Bennett comments:

In this passage (all of sec. 17) a complaint against a wrong analysis of subject concepts is jumbled together with a complaint against Locke's unsufficiently idealist analysis of reality.[16]

Bennett's judgment appears somewhat stretched here. Berkeley is merely contending that even if the concept of material substance as *substratum* (that in which the primary qualities inhered) made sense, it could not serve the purpose Locke wished it to have; but in fact, because the notion of "support" is unintelligible, the concept of material substance as substratum makes no sense. Bennett might respond, that if we were correct, Berkeley would have been willing, had he rejected his immaterialism (coming to believe that chairs, tables, etc., had an existence independent of minds), to make exactly the same criticism of the concept of "material substance" that he did as an idealist; the criticism that the notion of something underlying and "supporting" accidents is unintelligible. More precisely, Bennett would argue that if we were correct, Berkeley's critique of the concept of "matter" *was* (as opposed to *should* have been) considered by him to be independent of his immaterialism. And in support of the contention that Berkeley would *not* have accepted such a distinction, Bennett could refer to section 73 of the *Principles*, which more than the sections he explicitly mentions, gives evidence that Berkeley did conflate metaphysical issues with issues concerning the logic of the concept of "substance."

In that section Berkeley contends that it is because men believed that certain qualities (first "secondary" qualities like color, later "primary" qualities like extension) had an existence independent of minds, some

16 Bennett, *op. cit.*, p. 93.

"unthinking substratum" was required for these qualities to inhere in. For example, he suggests that when men realized that certain qualities like color were mind dependent, others such as extension and shape were considered to still exist without the mind, and therefore required a *"material support."* Berkeley concludes the section by remarking that since it has been demonstrated that *no* "quality" can exist unperceived, there "is no longer reason to suppose the being of "matter":

nay, that it is utterly impossible there should be any such thing so long as that word is taken to denote an unthinking substratum of qualities or accidents wherein they exist without the mind.

The section suggests that for Berkeley a material substratum *would* be required for qualities to inhere in, if the latter could have an existence independent of minds. This says no more than that, for Berkeley as for Locke, it is analytic of the concept of "qualities" that there be some "substance" in which they inhere.[17] There is no suggestion in this section that the concept of "matter" is unintelligible, *because* the alleged relation of "support" between "matter" and its accidents is unintelligible. Sections 73 through 76 can only be construed, not as a critique of the concept of "substance," but rather of the concept of an *"unthinking"* substance. Sections 73-76 of the *Principles,* however, are also compatible with the view that Berkeley utilized two independently sufficient conditions for finding the expression "material substance" devoid of significance, and, in these sections, was merely focusing on one of them.

Berkeley constrasts the "philosophic" sense of "material substance," as substratum, with what he considers to be a "common" sense ("vulagr" sense) of the expression, as "a combination of sensible qualities, such as extension, solidity, weight, and the like." (*Principles* 37) There is, however, a third sense to the expression, where it refers to the "insensible" or *in principle* unobservable particles of corpuscularian physical theory, particles which are alleged to have a causal role in the occurrence of the sensible properties of objects. This "sense" of the expression raises what we have called issue (2), the question of mind-body interaction, particularly as it is raised in the context of certain transmission theories of perception.

Transmission theories of perception are the subject of a good bit of discussion in the *Three Dialogues* (less so in the *Principles*), although the context in which these discussions appear varies considerably. (By

---

[17] What Berkeley does deny is that this "inherence" can be understood in terms of the material images of support or substratum. He allows the existence of spiritual substances. The relation of mind to its objects is not for him a relation of "support."

"transmission theory of perception" we understand a theory that attempts to explain sensation in terms of some material interaction – either impact or wave phenomena – between physical objects and the sensory and nervous systems of sentient beings.) In dealing with Berkeley's discussion of such theories, it is well to remember that there are a number of distinct problems he is concerned with, and it is not always clear which problem is the object of discussion. We distinguish at least four problem areas: (1) the problem of category mistake; (2) the problem of "causality"; (3) the problem of psycho-physical "laws" and (4) the problem of the existence of "insensible" particles.

Problem (1) is the subject of discussion early in the *Three Dialogues,* when the question of the mind dependence of sounds and colors is raised. Hylas remarks that "sounds" cannot "inhere" in "sonorous bodies," since a bell struck in a "receiver" exhausted of air will emit no sound, and he concludes that "air, therefore, must be thought the subject of sound." (*Dial.* 1-119.) Pressed as to whether "sound" can appropriately be said to be "in the air," Hylas contends that:

It is this very motion in the external air that produces in the mind the sensation of sound. For, striking on the drum of the ear, it causes a vibration which by the auditory nerves being communicated to the brain, the soul is thereupon affected with the sensation called "sound."

Berkeley (Philonous) points out that if "sound" were in truth the same thing as these "vibrative" or "undulatory" motions in the air, then it would be proper to predicate of motions qualities like "loud," "sweet," "acute," or "grave." This is a conclusion even Hylas is fundamentally unwilling to accept. The question here, however, concerns merely the proper use of "sound" predicates, and no questions are raised concerning the transmission or "causal" theory of sound perception enunciated by Hylas (*Dial.* 1-119-120).

Similar considerations apply to the discussion of color terms. Hylas appears to confuse a causal account of the occurrence of color sensations with the meaning of color terms. Again the rather detailed transmission theory of perception (put into the mouth of Hylas), in which the "minute particles" of light communicate "motions" to the optic nerve, and then to the brain, is not subject to analysis; Berkeley's point being that it is mistaken to identify "colors" (which are "ideas" or sensations) with "the motions and configurations of certain insensible particles of matter." (*Dial.* 1-125-126)

Early in the Second Dialogue, a transmission theory of perception *is* discussed in the context of the problem of mind-body interaction. The particular type of "interactionism" involved, however, is what we will term "macroscopic" interactionism, where the bodily or "material" side of the alleged causal relation consists of sensible as opposed to "insensible" objects. Hylas offers the Cartesian view of interactionism:

> It is supposed the soul makes her residence in some part of the brain, from which the nerves take their rise, and are then extended to all parts of the body; and that outward objects, by the different impressions they make on the organs of sense, communicate certain vibrative motions to the nerves and these, being filled with spirits, propagate them to the brain or seat of the soul, which, according to the various impressions or traces thereby made in the brain, is variously affected with ideas. (*Dial.* 2-150)

Hylas admits that the "brain" is itself a "sensible object," therefore a congeries of "ideas" which themselves (consistent with his earlier admissions) have no existence apart from minds. This leads Philonous to remark on the following apparent absurdity:

> Besides spirits, all that we know or conceive are our own ideas. When, therefore, you say all ideas are occasioned by impressions in the brain, do you conceive this brain or no? If you do, then you talk of *ideas* imprinted in an *idea,* causing that same *idea,* which is absurd. If you do not conceive it, you talk unintelligibly, instead of forming a reasonable hypothesis. (*Dial.* 2-151) (Our italics.)

The absurdity, however, is more apparent than real, and is rooted in Berkeley's equivocal use of his own technical term "idea." More precisely Berkeley fails to distinguish between an "idea" and the mental state of "being aware of an idea." For example, we can speak of a brain state (S) as the state of a "sensible object," and thus an "idea" (or collection of "ideas"), and we can also speak of the mental state (M) termed "the awareness of brain state (S)." It is also possible that there is a unique and invariable association between (M) and (S). We could formulate this association as a "psycho-physical" law that X's brain is in state (S) *only if* X is aware of his brain being in state (S). This would be a somewhat unique example (observation of one's own brain) of general laws or associations between brain states and mental states. (The latter would be expressed by locutions such as "awareness of the red patch," "experience of pain," "visual observation of one's brain," etc.)

Such "psycho-physical" laws are consistent with immaterialism, and there is evidence that Berkeley not only accepts, but makes use of them. We have discussed the theory of vision, where implicitly in the *Essay* and explicitly in *Visual Language* Berkeley makes use of correlations between the nature of the retinal image and certain characteristics of visual experience (e.g., "circle of confusion" and confused visual experience). It should be remarked that although retinal images and brain states are construed as "ideas" and therefore possessed by some mind, this is irrelevant to the formulation of laws associating them with mental states, just as it would be irrelevant in the formulation of laws that associate the states of one "sensible body" (e.g., lunar stages) with the states of another (height of the tides).

In discussing the distinction between divine and human cognition, Berkeley clearly indicates his acceptance of what we have termed "psychophysical" laws:

We are chained to a body; that is to say, our perceptions are connected with corporeal motions. By the law of our nature we are affected upon every alteration in the nervous parts of our sensible body; which sensible body, rightly considered, is nothing but a complexion of such qualities or ideas as have no existence distinct from being perceived by a mind; so that this connection of sensations with corporeal motions means no more than a correspondence in the order of nature between two sets of ideas, or things immediately perceivable. (*Dial.* 3-187) [18]

Berkeley's use here of the term "idea" is also somewhat equivocal, since it evidently refers both to the objects of consciousness and the consciousness of objects.[19] For example, my *awareness* of a red patch should be termed a "sensation," but not the patch itself, which in the Berkeleian idiom would be called a "sensible object" or "idea." Modes of conscious-

[18] Hylas argues, that if, in fact, the "ideas" we have are also somehow "in" or "known" by God, and pain is an idea, then God must suffer pain. The rejoinder of Philonous is clearly insufficient; he merely responds that God does not "perceive by sense." We can take that to mean that God does not have a "corporeal nature" (sense organs, nervous system, brain, etc.) and therefore is not "affected" as we are. Berkeley, however, himself contends that changes in our "corporeal nature" and our "sensations" are correlated contingently; there is no necessary connection between them. Therefore, Philonous' response says no more than that for God, one half of the correlation is absent, i.e., a "corporeal nature." But since such a nature is not necessarily connected with sensation, God might still suffer pain. In fact, Berkeley never explains in what sense God "knows" pain if he does not suffer it.

[19] The equivocation here is not unimportant, although a detailed analysis of its ground is beyond the scope of this essay. The fundamental problem is distinguishing between "*awareness* of an idea" (object of sense) and articulating the structure of an "idea." The latter requires no reference to consciousness. Berkeley, however, to emphasize his point that "ideas" cannot exist external to minds; that they are in minds, however metaphorically we interpret this "in," associates all "ideas" with what are more narrowly called "sensations," like "pain," or the feeling of warmth. It is a characteristic of such sensations that they appear to be "in" us, that is, modifications of our body. The "blueness" of the curtain, however, does not have the phenomenal character of being a modification of my body. One might say that the attribute of "consciousness" or "awareness" is built into the very concept of "pain," (and other sensations, certain tactile sensations, for example). Collapsing all objects of sense into this narrower class of "sensations," it is somewhat understandable, that Berkeley might consider even psychophysical laws as relations of "ideas."

ness, then, cannot be consistently called "ideas"; and it is such modes which form the second part of the correlations that constitute psycho-physical laws. Yet, even with this important distinction, psycho-physical laws conceived as regular associations between "bodily" states (where the body is viewed as a "sensible object") and mental states, are, as we have mentioned, accepted by Berkeley, and consistent with immateri-alism.

Berkeley's contention that it is *"absurd"* to speak of "ideas imprinted in an idea, *causing* that same idea," (our italics) can be viewed not as an objection to psycho-physical laws, but to the use of the term "cause." If we understand by "cause," efficient or productive causality, no "idea" can "cause" another "idea," since productive causality is reserved to designate the relation between will (as cause) and "idea" (as effect). From this point of view the correlation between the states of one type of "sensible object" and another cannot be considered a "causal" relation.

It is well to remember that Berkeley's fundamental critique of what he considers to be mistaken applications of the category of "efficient causality" is not uniquely related to the problem of "mind-body" inter-action. The traditional problem of such interaction has its roots in what we could call narrow or Cartesian mechanism; the view that the efficient or moving cause for the changes in motion of a material object is impact with another material object.[20] From this point of view it appears mys-terious how mental events (perceptions, images, reasoning, etc.) which are not extended, can "cause" changes in certain material objects, for example, the brain, nervous system, muscles, etc. What is not extended cannot "impact" with what is.

Since Berkeley disallows any efficient causality within the phenomenal world, this traditional problem disolves; it is as much (or as little) of a mystery how one billiard ball can move another through impact, as how a stone dropped in one's toe can produce a sensation of pain. Which is to say there is no mystery at all, since natural laws, either physical, or psycho-physical, exhibit no causal interaction, but merely regular associ-ation of different types of events.

Given Berkeley's view, there is some difficulty in interpreting some of his remarks concerning "mind-body" interaction. For example in the *Principles,* after discussing the difficulties in Locke's view, that though we are not directly aware of external objects, they can be said to have

---

[20] "Change in motion," would be more appropriate than "motion" since the principle of inertia asserts that a change in velocity, not merely a change in place requires a cause.

some characteristics (extension, figure, solidity) for which some of our "ideas" are "copies"; Berkeley remarks:

But though we might possibly have all our sensations without them, yet perhaps it may be thought easier to conceive and explain the manner of their production by supposing external bodies in their likeness rather than otherwise ... But neither can this be said, for, though we give the materialists their external bodies, they by their own confession are never the nearer knowing how our ideas are produced, since they own themselves unable to comprehend in what manner body can act upon spirit ... (*Principles* 19)

We might take the passage to mean that for Berkeley, a sufficient condition for denying the existence of "external bodies" ("matter" understood here not as substratum but as unperceivable entities describable in terms of the qualities of extension, figure, solidity and motion) is that it is incomprehensible how such bodies and "produce" sensations. If, however, "incomprehensible" refers to the impossibility of understanding how what is extended can effect any change in what is not, then it is difficult to see why Berkeley would take this problem seriously. As we have suggested, even in the realm of relations among macroscopic "physical" objects (for example, the laws of impact) there is no effecient causality.

A rejoinder to these comments might focus on our term "macroscopic," by which we understand what Berkeley calls a "sensible object." It might go as follows: Macroscopic objects are no more than collections of "ideas"; "ideas" are "passive," and therefore relations among such objects exhibit no efficient causality. Even impact phenomena can be construed as sequences of perceptual states in which no causal agency is perceived.[21] Such a view makes Berkeley's denial of efficient causality within the natural world a consequence of immaterialism, implying perhaps that had he rejected immaterialism, he would have allowed efficient causation within the material world, and hence been faced with the dilemma of how material change in the brain and nervous system can "cause" sensation.

On the other hand, it would hardly seem likely that even if Berkeley had accepted the existence of "insensible particles" independent of minds, he would have attributed to them causal efficacy, a property he reserves for will. As we will see when we discuss *De Motu*, immaterialism, with his attendant doctrine of the "passivity" of ideas, no longer has a crucial role in the denial of efficient causation within nature; more important is Berkeley's belief that the attribution of causal efficacy has no explanatory function in the formulation of natural laws (laws of impact; law

[21] See *De Motu*, 22, 26.

of gravity). There is no reason Berkeley would not include psychophysical laws within the class of "natural" laws. Hence, even if he rejected immaterialism, Berkeley would view the alleged dilemma of how matter *affects* mind as a falsely posed one, a misuse of the category of causality.[22]

Perhaps the best view, then, of *Principles* 19, is not that Berkeley takes seriously the problem of comprehending a causal relation between "matter" (understood again as particular "insensible" bodies) and "mind" (understood as "mental events," or individual acts of perception, imagination, reasoning, emotion, etc.), but that he is reminding the "materialists" that if they accept (a) that efficient causality in the material world can only ultimately be understood in terms of the impact of one material object on another and (b) that mental events (for example, sensations) are not material objects; they will never comprehend how a change in the state of a material object ("brain") could cause a sensation (the experience of pain).

In section 50 of the *Principles,* Berkeley attempts to refute what he understands to be an objection to immaterialism: that advances in science (the "study of nature") presuppose the existence of "corporeal substance or matter."

To this I answer that there is not any one phenomena explained on that supposition which may not as well be explained without it, as might easily me made appear by an induction of particulars. To explain the phenomena is all one as to show why, upon such and such occasions, we are affected with such and such ideas. But how matter should operate on a spirit, or produce any idea in it, is what no philosopher will pretend to explain; it is therefore evident there can be no use of matter in natural philosophy. Besides, they who attempt to account for things do it not by corporeal substance, but by figure, motion, and other qualities, which are in truth no more than mere ideas and, therefore cannot be the cause of anything.

The passage is unfortunately ambiguous on a number of important issues. In the first place, it is unclear what is being referred to by the expression

[22] The point can be further illustrated by looking at the following passage from the *Siris* (251).

"The Democritic hypothesis, saith Dr. Cudworth (*True Intellectual System* 1678) doth much more handsomely and intelligibly solve the phenomena than that of Aristotle and Plato. But, Things rightly considered, perhaps it will be found not to solve any phenomenon at all; for all phenomena are, to speak truly, appearances in the soul or mind; and it hath never been explained, nor can it be explained, how external bodies, figures and motions, should produce an appearance in the mind. Those principles, therefore, do not solve, if by solving *is meant* assigning the real, either efficient or final cause of appearance, but only reduce them to general rules." (Our emphasis) The passage is misleading however since it suggests that the problem is how material changes can be the efficient causes of mental states. For Berkeley, however, material changes could not be said to be the cause even of other material changes.

"corporeal substance or matter." Is it "material substance" as substratum or "support" of the primary qualities? Or is it "material substance" as "insensible bodies" (Newton's "hard massy particles") [23] posited in atomic theories? One would guess the former reference is meant, since later in the passage Berkeley remarks that those who attempt to explain phenomena, do not do so by "corporeal substance," but "by figure, motion, and other qualities..." And if we are right, there is no difficulty in agreeing with Berkeley that the concept of "corporeal substance" has no role in the progress of physics.

Secondly, what is meant by "explaining" a phenomenon by "an induction of particulars?" One might think it is to exhibit the phenomenon in question as an example or particular instance of a more general law (or regularity). We have a specific illustration of this in the discussion of "gravity" later in the *Principles*:

> For example, in the falling of a stone to the ground, in the rising of the sea toward the moon, in cohesion and crystallization, there is something alike, namely a union or mutual approach of bodies... Thus he (a philosopher) explains the tides by the attraction of the terraqueous globe toward the moon which to him does not appear odd or anomalous, but only particular example of a general rule or law of nature. (*Principles* 104) [24]

It is unclear from this particular passage whether the expression "general rule" refers to the claim that "bodies have a mutual tendency towards each other," ("denoted by the general name 'attraction' ") or the mathematical law of gravity, of which the qualitatively diverse types of "attraction" (e.g., tidal phenomena, planetary orbits, free fall) are, when quantified, deducible as consequence. Textual considerations, particularly attention to *De Motu* and the *Siris,* indicate Berkeley means the latter, and does not mean that simple inductive generalizations can be

---

[23] Newton, *Optics, op. cit.,* p. 400. "All these things being considered, it seems probably to me, that God in the Beginning form'd matter in solid, massy, hard, impentrable movable particles, of such Sizes and Figures, and with such other properties, and in such Proportion to Space, as most conducted to the End for which He form'd them ..."

[24] For example, *Siris* (245). "The ancients had some general conception of attracting and repelling powers as natural principles ... But Sir Isaac Newton, by his singular penetration, profound knowledge in geometry and mechanics, and great exactness in experiments, hath cast a new light on natural science. The laws of attraction and repulsion were in many instances discovered, and first discovered by him ..."

Berkeley's point here, and in other passages, is that Newton made more progress than the Cartesians and others ("corpuscularians") who attempted to explain gravitational phenomena. Berkeley makes clear that "progress" does not refer to understanding more of the "efficient cause" of such phenomena. What is significant is that Berkeley accepts in the *Siris* principles (laws) of attraction and repulsion as fundamental principles of mechanics, needing no reduction to the laws of impact.

said to "explain" their instances. Formal considerations might indicate the same conclusion. Limiting ourselves to what Berkeley calls "mechanical explanation" (that is to "explain" is, at least in part, to deduce the phenomena from accepted and more general laws) we would not wish to say that the inductive generalization "all bodies have a mutual tendency towards each other" "explains" why this pair of bodies has a mutual tendency towards each other, although the purported *explanans* is (in terms of extension) more general than the *explanandum,* and the latter *is* deducible from the former.[25] With respect to "gravity," then, we would interpret "induction from particulars" (Sec. 50 *Principles*) to mean more than generalizing from the observed cases where bodies have a mutual tendency towards one another, but to mean reducing that "tendency" to a "rule" or mathematical law. We might add, that if Berkeley meant to adduce as an explanation for the observed joint occurrence of (A) and (B), the simple inductive generalization, "All (A) are (B)," his contention in the *Principles* (sec. 50) that

to explain the phenomena is all one as to show *why,* upon such and such occasions, we are affected with such and such ideas, (our italics)

would make little sense. If we reject a claim of efficient causality as a legitimate answer to the "why"; what type of answer would be satisfactory? The generalization that "upon such and such occasions we are *always* affected with such and such ideas," would hardly seem adequate, since it is the correlation itself which demands explanation, and this demand is not served through introduction of the quantifier "all" or "every." Berkeley, as far as we know, does not explicitly distinguish between explaining an individual event and explaining a "law" (or regular

---

[25] Modern discussions have gone into a good deal more detail concerning this "deductive-nomological" model of explanation; for example, distinguishing formal from empirical conditions of adequacy, distinguishing explaining individual events, from explaining laws, etc. A good discussion is found in Ernst Nagel, *The Structure of Science,* (New York: Harcourt, Brace and World, Inc., 1961) Chapters 3 and 4. Berkeley's use of the model is consciously patterned after the methodological discussions of Newton, particularly in the *Principia.* For example the latter's well known statement of method in the Preface to the First Edition of the *Principia,* "for the whole burden of philosophy seems to consist in this – from the phenomena of motions to investigate the forces of nature, and from these forces to demonstrate the other phenomena." (Isaac Newton, *Mathematical Principles of Natural Philosophy,* trans. Andrew Motte, revised Florian Cajori, (Chicago: Encyclopedia Britannica, 1952) p. 1. See Berkeley's second letter (Nov. 25, 1729) to his American correspondent, Samuel Johnson; section 1. "The true use and end of natural philosophy is to explain the phenomena of nature, which is done by discovering the laws of nature, and reducing particular appearances to them. This is Sir Isaac Newton's method; and such method or design is not in the least inconsistent with the principles I lay down." (Berkeley, *Correspondence,* Ed. Turbayne, p. 224)

association between types of events), but is would seem likely that it is the latter that is referred to in the above passage; an observed regular association of types of events is "explained" when it is shown to be a consequence of more general laws.

Lastly, we would raise some questions about Berkeley's phrase that

they who attempt to account for things do it not by corporeal substance, but by figure, motion, and other qualities, which are in truth no more than mere ideas and therefore, cannot be the cause of anything, as has already been shown. (*Principles* – 50)

The passage can be construed, as we have seen, as a critique of the concept of material substance as the "substratum" or "support" of the primary qualities. Since Berkeley has claimed the concept is vacuous, it would be meaningless to attribute to "corporeal substance" any causal efficacy.

The passage, however, can also be viewed as a claim that scientific laws, both physical and psysical and psycho-physical, deal with the regular associations between "sensible" or observable phenomena; where "observable" has the broad sense of what is apprehended through the senses, or by reflection on our own mental life (in Locke's phrase "perception of the operations of our own mind within us").[26]

Such a view of the passage implies that Berkeley is contending, in effect, that although the "primary qualities," mass and velocity, for example, play a crucial role in the formulation of the laws of mechanics (e.g., the laws of impact and the law of gravitation), such qualities must be considered to be observable, hence "ideas." In the same way, psychophysical laws which correlate bodily changes with states of consciousness are legitimate only when the "body" and its parts are considered "sensible objects." It would follow that Berkeley not only denies the existence of "material substance" viewed as "substratum" for the qualities that enter into the laws of mechanics, but denies, as well, the existence of those "insensible particles" posited in atomic or corpuscularian theories of nature.

It must be remembered that the grounds for the claim of "insensibleness" with respect to such particles is quite different from the grounds for such a claim with respect to "material substance" as "substratum"; a point emphasized by Bennett.[27] Material substance as substratum is "insensible" simply because the concept of "material substance" (in this

---

[26] Locke, *Essay*, Book 2 Chapter 1 Paragraph 4.
[27] Bennett, *op. cit.*, pp. 91-104.

sense) is vacuous. One can form no "idea" of anything that would fall under a self-contradictory or vacuous concept.

That the alleged "particles" of atomic or corpuscularian theories are "insensible," is entailed by a certain version of what can be called the "causal theory of perception." [28] This version runs essentially as follows: There exist real (material) objects independent of minds. The nature ("real essence" in Locke's terminology) of such objects is construed as a certain "atomic" structure, by which is meant a certain combination of "particles" whose nature is exhaustively described in terms of qualities like shape, motion, mass (or the less precise term "solidity"). Other qualities might be added as essential, such as a certain *vis inertia,* and attractive (or repulsive) power. There is a good deal of historical debate about whether these latter qualities are "essential," but we need not go into that here. What is important is that these alleged particles are said to interact by certain modes of transmission with the nervous and sensory systems of sentient beings (these latter construed also as certain groupings of these particles). The modes of interaction between the "object" and the physiological systems would ideally be described in terms of mechanical laws, viewing "mechanism" here in a broad enough sense to allow claims that the particles have certain irreducible attractive or repulsive powers. The end result of such interaction is sensation, or the having of "ideas." Although the spatial terminology is somewhat ambiguous and metaphorical, we are to understand that the *locus* of the sensation is "in" the sentient being. Expressed otherwise, we are immediately acquainted not with the external objects which are causally related to the occurrence of our "ideas" but only with the "ideas" caused.

In is a consequence of the above theory that the ultimate "particles" which constitute the atomic structure of a physical object, are in principle unobservable, an unobservability certainly not rooted in some empirical limitation of our powers of perceptual discrimination, but rooted in the claim that the content of a perceptual consciousness (in terms of the colors, sounds, tastes, feels, smells) is an effect of the interaction of the interaction of the particles of "objects" with the particles of our sensory and nervous systems. These ultimate "particles" are then "theoretical" entities, understanding "theoretical" in what we will term a "strong

---

[28] There can be other versions then the one we mention in the text. The causal theory of perception claims that external objects are causally related to our perception of them. This is compatible with some form of "realism" which contends that we are at times (if not invariably) immediately aware of the qualities of the object itself. A causal theory of perception does not logically entail a "representative" theory of perception.

THE PHILOSOPHY OF PHYSICS

sense." An entity is said here to be "theoretical" if (a) it functions in a theory, in this case a theory describing the "material" conditions for perception, and (b) it is in principle unobservable.[29]

The question before us is whether Berkeley would deny not only the existence of such "theoretical" entities, but would also deny their usefulness (as "hypothetical" entities) in the formulation of scientific laws. Passages from the *Principles* and the *Three Dialogues* where "corpuscularian" theories are discussed are not necessarily helpful here, since Berkeley's critical attention is often directed at the contention that the "insensible particles" are the *efficient cause* of our "ideas" (that is, that they "produce" our ideas). *Principles* (102) is an excellent example of this:

> One great inducement to our pronouncing ourselves ignorant of the nature of things is the current opinion that everything includes within itself the cause of its properties; or that there is in each object an inward essence which is the source whence its discernible qualities flow, and whereon they depend. Some have pretended to account for appearances by occult qualities, but of late they are mostly resolved into mechanical causes, to wit, the figure, motion, weight, and suchlike qualities of insensible particles; whereas in truth there is no other or efficient cause than spirit, it being evident that motion, as well as all other ideas is perfectly inert. Hence to endeavor to explain the production of colors or sounds by figure, motion, magnitude and the like, must needs be labor in vain. And accordingly we see the attempts of that kind are not at all satisfactory.

It may be true that Berkeley's fundamental concern with the issue of "efficient causation" obscured for him the role that "theoretical entities" might play in the scientific explanation of "macroscopic" or "observable" phenomena. And it is certainly true that in Berkeley's time the "atomic" theory of matter had not achieved any of the astounding successes it was later to achieve in explaining [30] and unifying, for example, the classical laws of thermodynamics, to say nothing of its later success in explaining

---

[29] A "weak" condition for (x) to be a "theoretical entity" would be merely that (x) functions in a "theory" without claiming that (x) is in principle unobservable. In what might be considered the "classic" discussion of what it means to be defined by a theory; two fundamental aspects are involved: (a) the entity is formally defined by the postulates of the theory; (b) by means of "bridge principles" or "coordinating definitions" the entity formally defined, or some "state" of the entity formally defined; (i.e., the "momentum" of a molecule) is said to correspond to something observable. A classic example used by Norman Campbell is the theoretical entity "molecule" as it functions in the kinetic theory of gasses. (Norman Campbell, *Foundations of Science*, (originally, *Physics the Elements*, Cambridge University Press, 1920) (Reprinted; New York: Dover Publications Inc., 1957) Chapter 6.

[30] We understand "explain" here, as Berkeley would; deducing certain observable regularities from more general laws; only in this case the latter would make use of "unobservable" entities.

electrical, magnetic and chemical phenomena. On the other hand, speculations about the particulate character of "light" had been developed in
some detail by Newton, and as we will see below, were well known to
Berkeley.

Karl Popper in a generally excellent monograph on Berkeley's Philosophy of Science contends that Berkeley's "phenomenalism" obscured his
understanding of the role of theoretical entities in scientific explanation.[31]
It is true that, strictly speaking, "immaterialism" entails that unobservable
shapes, motions, and masses could not exist. However, it would not
preclude entities defined in terms of such properties functioning as conceptual aids in unifying, in one theoretical framework, diverse kinds of
phenomena.

Popper is more accurate, we believe, in nothing that Berkeley rejects
what he (Popper) calls "essentialist or metaphysical causal explanations." [32] For Berkeley the fundamental mistake in the causal theory
of perception *is* its attribution of causal efficacy to the "insensible particles" of corpuscularian theories, rather than that such theories posit the
existence of unobservable entities.

To buttress his view that Berkeley is suspicious of if not hostile to
speculative hypotheses that go beyond what is observable, Popper quotes
part of a very interesting passage from the *Siris*.

It is one thing to arrive at general laws of nature from a contemplation of
the phenomena; and another to frame a hypothesis, and from thence deduce
the phenomena. Those who suppose epicycles, and by them explain the
motions and appearance of the planets, may not therefore be thought to
have discovered principles true in fact and in nature. And, albeit we may
from the premises infer a conclusion, it will not follow that we can argue
reciprocally, and from the conclusion infer the premises. For instance, supposing an elastic fluid, whose constituent minute particles are equidistant
from each other, and of equal densities and diameters, and recede one from
another with a centrifugal force which is inversely as the distance of the
centers; and admitting that from such supposition it must follow that the
density and elastic force of such fluid are in the inverse proportion of the
space it occupies when compressed by any force; yet we cannot reciprocally
infer that a fluid endowed with this property must therefore consist of such
supposed equal particles. (*Siris* 228) [33]

Popper views the passage as evidence that Berkeley, on the one hand,
astutely recognizes that explanatory hypotheses are not logically entailed
by the phenomena they explain; and on the other hand, falsely concludes

[31] Popper, *op. cit.*
[32] *Ibid.,* p. 443.
[33] *Ibid.,* p. 449.

that such hypotheses are to be rejected, at least if they pass beyond generalizing the observed relationships between "sensible" bodies.

In our view, however, Popper seriously misreads the passage. The full context of the passage makes clear that what Berkeley is criticizing is the use of what can be called "ad-hoc hypotheses," whose sole purpose is to explain certain phenomena; and where there is either no more direct evidence for the hypothesis than the phenomena to be explained (no evidence independent of what went into the construction of the hypothesis), or, more seriously, when there is direct evidence that the hypothesis is false (phenomena logically incompatible with the hypothesis). The remainder of the passage referred to by Popper reads as follows:

for it would then follow that the constituent particles of air were of equal densities and diameters; whereas it is certain that air is a heterogeneous mass, containing in its composition an infinite variety, of exhalations, from the different bodies which make up this terraqueous globe. (*Siris* 228)

Siris 228 when conjoined with other passages indicates that what Berkeley is criticizing are hypotheses like the Cartesian vortex theory to explain gravitational phenomena. For example, we find in *Siris* (232)

Nothing could be more vain and imaginary than to suppose with Descartes that merely from a circular motion's being impressed by the supreme agent on the particles of extended substance, the whole world, with all its severall parts, appurtenances and phenomena might be produced by a necessary consequence from the laws of motion.

In fact, in the above and other passages it seems apparent that Berkeley is doing no more than echoing Newton's critique of the speculative hypotheses of Cartesianism, a critique summed up the famous expression of the *Principia*, "I feign no hypotheses." [34] Newton certainly would have recognized that his own gravitational law was not logically entailed by certain relationships among phenomena (the relation, for example, between distance and time fallen for a body in free fall in the earth's gravitational field) "explained" by his law. Moreover, Newton's critique of Descartes' vortex theory indicates that he could have found such a theory acceptable, if, in fact, it had squared with observed fact (for example, Kepeler's law relating the period of a planet's orbit around the sun to its mean distance from the sun).[35]

[34] Newton *Principia, op. cit.,* "General Scholium" to the second edition, p. 371. We accept the translation of "fingo" as "feign" not "frame" by Alexander Koyre, *Newtonian Studies,* (Chicago: University of Chicago Press; 1968 – first published – 1965) p. 25.
[35] See Newton, *Principia, op. cit.,* "General Scholium" at the end of Book III, p. 369. "The hypothesis of vortices is pressed with many difficulties. That every planet

Our own reading of the *Siris* suggests that Berkeley's critique of Cartesianism presented there is based neither on the fact that such hypotheses posit the existence of "insensible particles," nor on the fact that such speculative hypotheses are not logically entailed by the phenomena they purport to explain, but rather on the fact that such hypotheses, like the Ptolmaic "epicycles," have the character of being "ad hoc," that is framed without any serious consideration as to whether they are in fact true.

The strongest support for our view comes from the *Siris* itself, where Berkeley appears to entertain (and perhaps accept) speculative corpuscular hypotheses concerning the nature of the "aether" and/or of "light"; hypotheses modeled on the suggestions in Newton's *Optics*, particularly certain "queries" attached to the end of that work.[36] Section 162 of the *Siris* (there are many others that are comparable) is a good example of Berkeley's speculative endeavors:

The pure aether (which Berkeley identifies with the first "corporeal" "instrument" of the divine will) or invisible fire contains parts of different kinds, that are impressed with different forces, or subjected to different laws of motion, attraction, repulsion and expansion, and endued with diverse habitudes toward other bodies ... the different modes of cohesion, attraction, repulsion, and motion appear to be the source from whence the specific properties are derived, rather than different shapes of figures.

This passage, again, offers a criticism of the Cartesian view that physical phenomena could all be explained in terms of the modes of extension alone. In fact Berkeley chides Newton for suggesting in the *Optics* that gravitational phenomena could be explained in terms of variations in the density of the aether.[37] Of importance for us, however, is Berkeley's contention that the "aether" (or "light" which Berkeley argues is "corporeal")[38] is "invisible," known only by its "effects." Later in the work this view is again expressed:

by a radius drawn to the sun may describe areas proportional to the times of description, the periodic times of the several parts of the vortices should observe the square of their distances from the sun; but that the periodic times of the planets may obtain the 3/2th power of their distances from the sun, the periodic times of the parts of the vortex ought to be as the 3/2th power of their distances.

[36] Newton, *Optics, op. cit.,* Query 18, 28, and 31 are referred to by Berkeley, as well as other passages in the *Optics*.

[37] *Siris* (224).

[38] *Siris* (207), "But it is now well known that light moves; that its motion is not instantaneous; that it is capable of condensation, rare-faction, and collision; that it can be mixed with other bodies, enter their composition, and increase their weight ... all of which seem sufficiently to overthrow those arguments of Ficinus, and show light to be corporeal."

As in the microcosm, the constant regular tenor of the motions of the viscera and contained juices doth not hinder voluntary motions to be impressed by the mind on the animal spirit; even so, in the mundane system, the steady observance of certain laws of nature, in the grosser masses and more conspicuous motions, doth not hinder, but a voluntary agent may sometimes communicate particular impressions to the fine aetherial medium, which in the world answers to the animal spirit in man. Which two (if they are two), although invisible and inconceivably small, yet seem the real latent springs whereby all the parts of this visible world are moved-albeit they are not to be regarded as a true cause, but only as an instrument of motion, and the instrument not as a help to the Creator, but only as a sign to the creature. (*Siris* 261)

The viewpoint is fundamentally that of Newton in the "queries" appended to his *Optics:* "Sensible" phenomena are to be explained in terms of the interaction of ultimate (indivisible) particles, particles characterized by a *vis i :ertia* and moved by certain "active Principles, such as that of Gravity, and that which causes Fermentation and the Cohesion of bodies." [39]

Berkeley stresses, however, as he does in *De Motu,* that Newton's "active principles," while distinct from Cartesian principles of mechanics, are not to be construed as efficient causes. Principles such as "gravity," "magnetism" and so forth are elliptical concepts for certain forms of motion between the particles.

We are not therefore seriously to suppose with certain mechanic philosophers, that the minute particles of bodies have real forces or powers, by which they act on each other, to produce the various phenomena in nature. The minute corpuscles are impelled and directed, that is to say moved to and fro from each other, according to various rules or laws of motion. The laws of gravity, magnetism, and electricity are diverse." (*Siris* 235)
The words "attraction" and "repulsion" may, in compliance with custom, be used, where, accurately speaking, motion alone is meant. (*Siris* 240)

There is some question of how compatible the substantive scientific speculations in the *Siris* are with the more epistemologically oriented earlier works. Certainly Popper is mistaken in his suggestion that Berkeley

eschewed corpuscularian hypotheses. Even in the earlier works on vision they play a role.

T. E. Jessop, editor of the *Siris,* claims that the work is compatible with the "immaterialism" of the earlier works, since the qualities of the "particles," such as extension, weight, motions, which enter into the formulation of scientific laws are "ideas." [40] And one can refer, in support

---

[39] Newton, *Optics, op. cit.,* particularly "Query 31."
[40] *Siris,* p. 85, footnote by Jessop.

of this view, to passages as *Principles* (50) where Berkeley, as we have mentioned, remarks:

Besides, they who attempt to account for things do it not by corporeal substance, but by figure, motion, and other qualities, which are in truth no more than mere ideas and therefore, cannot be the cause of anything ...

However, *Principles* (50) could be viewed as dealing with "macroscopic" laws; those, for example, of gravitational astronomy, or classical thermodynamics. Moreover Jessop's view comes up against what appears to be an insuperable obstacle: the clear suggestion in the *Siris* that the "aether" (or "light" or first "corporeal" "principle") is "insensible," known only by its effects. In what sense can this "aether" (or its constituent particles) be called "ideas?"

Our own reading of the *Siris* suggests that Berkeley has left us with not so much a straightforward contradiction with the earlier work, but with an unresolved dilemma. Consistent with the earlier work, including *De Motu*, Berkeley insists in the *Siris* that the particles of the aether are not to have attributed to them causal efficacy; they are, he suggests not causally required by the "Creator," but rather a "sign" for the "creature." (*Siris* 261) Yet, if the particles are "insensible," in what sense can they be considered "signs" if by this terms we mean an "idea" separable from and suggestive of another "idea?"

The dilemma might be resolved if we interpreted "sign" broadly enought to mean, that which enables us to predict (or retrodict) sensible phenomena. In this view, corpuscular theories themselves, as opposed to an individual theoretical entity, could be called "signs," or instruments for prediction. Such a view is compatible with Berkeley's broad conception of the purpose of science: not to reveal the "essences" of things, but rather to extend our foresight and practical mastery of nature. Such a view of "signs" combined with an instrumentalist view of theoretical entities – viewing them not as "existing" but as useful devices (like the "lines and angles" of geometers) [41] – would allow Berkeley to entertain corpuscular hypotheses without violating his immaterialist metaphysics. Unfortunately, if we take the point of view of the texts themselves, particularly the *Siris*, there is, in our judgment, no convincing evidence that Berkeley took such an instrumentalist view. Contra Warnock [42] there is as well no evidence that Berkeley did, or would treat the ultimate particles of the "aether" as a "mathematical hypothesis," as he would so treat

[41] *Essay on Vision* (12).
[42] Warnock, *op. cit.*, p. 203.

"gravitational force." And the reason for this, from a formal point of view, is quite simple. The "insensible" particles are alleged to have certain properties like mass, shape, motility, and these are properties alleged to be distinct from their effects, although it may be true they can only be inferred from their effects. The issues concerning "force," as we will see when we discuss *De Motu*, are quite distinct. "Force" as used improperly (from Berkeley's point of view) by those who posit an immanent and animating principle of motion in material bodies, although construed as an "entity" is not even alleged to have any properties distinct from its "effects." And, as Berkeley points out, this is precisely one of the problems with the concept of force as an "occult" quality, that it is a vacuous concept.

Before turning to *De Motu*, and the concept of "force," there is one remaining issue concerning the concept of a "matter" which we will discuss. In a letter to Samuel Johnson, Berkeley raises a problem which occupied him from his early jottings in the *Philosophic Commentaries* to his last major work, the *Siris*.[43] Berkeley writes:

It must be owned, indeed, that the mechanical philosophers do suppose (though unnecessarily) the being of matter. They do even pretend to demonstrate that matter is proportional to gravity, which if they could, this indeed would furnish an unanswerable objection. But let us examine their demonstration ... (Second letter sec. 1)

It might be asked, what concept of "matter" is being referred to here? Philosophically it would seem to be "matter" as substratum or insensible "support" for the primary qualities. Yet Berkeley apparently chooses here to identify it conceptually with "mass," or more precisely with "inertial mass" as distinct from "weight" or "gravitational mass," what Berkeley calls "gravity." Most likely Berkeley's reference is to Newton, for it is the latter who distinguished the invariant character of inertial mass from the variability of weight, and claimed to demonstrate through pendulum experiments that "mass" (inertial mass) is proportional to "weight" (gravitational mass).[44]

---

[43] For example, in the *Siris*, Berkeley contends that to explain gravitational phenomena in terms of the variable density of the aether is circular since the density of one particle of aether would have to be defined in terms of its "gravity" (weight). The more general problem, is that if "mass" is to be defined in terms of density, and to be defined independently of weight; then density must be defined independently of weight. Against the claim that Newton's definition of "mass" is circular, there is the contention of Bernard Cohen, that for Newton "density" could be and was defined in terms of "specific gravity." I, Bernard Cohen, "Isaac Newton's Principia, The Scripture, and Divine Providence," (notes to appendix 1) (*Essays in Honor of Ernst Nagel*, ed. Morgenbesser, Suppes, White) (New York: St. Martin's Press; 1969) p. 541.

[44] Newton, *Principia, op. cit.*, Book 1, Def. 1.

Berkeley's contention is that the alleged demonstration of the proportionality between "matter" (Newton's "mass") and "gravity" (weight) commits a *petitio principi*. His argument is as follows:

(a) Momentum is said to be the product of velocity by quantity of matter. (MV)

(b) If velocity is given, momentum is proportional to the quantity of matter.

$$\frac{M_1V}{M_2V} = \frac{M_1}{M_2}$$

(c) Bodies descend *in vacuo* with the same "velocity" (Berkeley probably means *acceleration* here.)

(d) Therefore it is alleged that the "momentum of descending bodies" (which Berkeley calls "gravity") "is as the quantity of moles," or that "gravity is as matter." Symbolically if the "gravity" or weight of a body of Mass (1) is M_1a and the weight of a body of Mass (2) is M_2a; where a = acceleration in vacuo; then

$$\frac{M_1}{M_2} = \frac{W_1}{W_2}$$

The circularity occurs if you answer the question, how "estimate" the "moles" or "quantity of matter?" by saying "weight," producing the tautology that the quantity of matter is proportional to itself.

The historical problem was to gain an operational definition of "inertial mass" independent of "weight." In definition I of the *Principia*, Newton defines "mass" ("quantity of matter") as "the measure of the same, arising from its density and bulk conjointly." This suggests a circularity, if "density" itself is defined as the "mass" per unit volume.[45] In the work of Ernst Mach we do get an operational definition ("sensible measure") of the ratio of the masses of two bodies in terms of the negative ratio of their mutually induced accelerations; a definition independent of the concept of weight.[46] This definition, we think, would have been acceptable to Berkeley, and he would have contended undoubtedly that we mean no more by "mass" than this "sensible measure." Accelerations, as motions, could be construed as "ideas," and therefore, as far as physics is concerned, the concept of "mass" would not require any break with immaterialism.

From this point of view it is difficult to understand Berkeley's claim that if the proportionality between "matter" and "gravity" could be

---

[45] See the discussion by I. Cohen *loc. cit.*

[46] Ernst Mach, *The Science of Mechanics,* sixth ed. (first German edition 1883) translation, Thomas J. McCormack, (La Salle I 11.: Open Court Publishing Co; 1960) p. 238.

demonstrated, philosophers would be furnished with "an unanswerable objection." We take this to mean that they (philosophers) would be furnished with evidence for the existence of "matter." Here Berkeley does appear to confuse his logical critique of the concept of "material substance" with the scientific problem of clarifying the concept of "inertial mass" as distinct from weight. As Bennett has suggested, the force of the critique of "substance" if properly understood is that in principle there could be no sensible property labeled "substantiality." For that property (qua property) would have to be instantiated in some substance.[47] One might remark that the property of solidity (for Locke a "primary quality") is that "sensible" mark of "substance" which when clarified conceptually becomes the property of "inertial mass." It is the invariancy of inertial mass in the Newtonian system which gives operational meaning to our intuition that in material bodies something remains constant.

Aside from sensation of solidity, there was a good deal of evidence available to Newton concerning the "inertial" properties of bodies (resistance to changes of motion), properties independent of manifestations of weight.[48] The evidence must have been known to Berkeley. And, as in the *Essay on Vision,* Berkeley finds no difficulty in allowing for a certain invariancy to length, viewed as a metric property, he would (or should) have no difficulty in allowing for an empirically determined invariancy of something called "mass." Neither Mach's definition nor any comparable one could be the "unanswerable objection" to immaterialism.

The concept of inertial mass might, of course, cause some difficulties for Berkeley, for "inertia," or the property of resisting changes in motion, was considered by Newton to be an inherent as opposed to a relational property of bodies.[49] This suggests that is part of the nature of bodies to have the "power" of resisting changes in motion; in other words all bodies have an internal ("innate") force, (vis inertiae) which, for Newton, was a dynamic way of considering the "mass" or "quantity of matter" of a body. Newton held that this power of resisting changes in motion was a dispositional property of bodies manifested only in the face of some

---

[47] Bennett, *op. cit.*

[48] Mach, *op cit., p.* 237-240. Mach indicates some of the empirical evidence available to Newton that showed the "inertial" properties of bodies (their resistance to changes in motion) are distinct from manifestations of weight. This evidence of something called "mass" as distinct from weight must have been known to Berkeley, and therefore his apparent contention that "mass" must be defined in terms of weight ("gravity") is difficult to understand. Moreover, to allow "inertial" phenomena as distinct from "gravitational" phenomena, would not seem to be an obstacle to "immaterialism."

[49] Newton, *Principia, op. cit.,* Book I, Definition III.

"impressed" (external) force; yet it also was an inherent property, in the sense that if there was only one body in the universe it would have such a property.

Since a good deal of Berkeley's *De Motu* concerns the use of the term "force" in mechanics, we would expect that the principle of inertia would come in for some consideration. It does, and we will discuss Berkeley's comments on the principle in the next section. We will argue there that Berkeley's failure to come to grips with the meaning of the principle of inertia makes his attempt to deny to alleged "forces" (such as gravitational force) the status of efficient causes, ultimately unsuccessful.

B. DE MOTU: THE CONCEPT OF FORCE

There is an ambiguity in Berkeley's discussion of "force," but an ambiguity rooted, as we will see, in the equivocal character of the concept itself, as its meaning was elaborated in the development of mechanics. On the one hand "forces" and "force effects" are considered to be at least conceptually, if not referentially, separable. That is, "forces" are invoked to explain the occurrence of certain phenomena which are considered to need explaining; the deformation of material bodies, for example, or in the case of Newtonian mechanics, non-uniform motion.[50] In this sense "force" has the status of "cause" or principle of explanation. Berkeley's language often suggests that he in fact makes this conceptual separation. For example in *De Motu*, section 10:

We must however admit that no force is immediately felt by itself, nor known to measured otherwise than by its effect.

Or section 11:

Then no force makes itself known except through action and through action it is measured.

On the other hand "forces" such as "gravitational" or "magnetic" or "electrical" "forces" are not sensibly distinct from their effects, whether these latter are viewed in terms of material deformations, of the acceleration of bodies, or both. As a consequence Berkeley would plausibly find it difficult to assign to such "forces" the status of "causes" even given the

[50] Our discussion has been influenced by the suggestive article, "The Origin and Nature of Newton's Laws of Motion," by Brian Ellis; printed in *Beyond the Edge of Certainty, Essays in Contemporary Science and Philosophy* ed. Robert Colodny, (New Jersey: Prentice Hall, 1965) pp. 29-68. A lengthy critique of Ellis's article is found in *Philosophy of Science* Vol. 36, No. 3, September 1969; "Force and Natural Motion." I. E. Hunt, W. A. Suchting; pp. 233-251.

conception of causality developed in the *Principles* and elsewhere: that the relation "(A) causes (B)" is to be interpreted as a relation between sensibly distinct phenomena (or kinds of phenomena) in which (A) functions as the "sign" of (B). With impact phenomena, at least, one can sensibly isolate the "impact" or collision from the consequent motion, allowing such phenomena to be incorporated within the Berkeleian concept of "cause." This is not the case with gravitational or magnetic phenomena, to take merely two examples.

This double character of certain "forces" – that they function, on the one hand, as causal or explanatory principles, and are not, on the other hand, sensibly isolable from their effects – is the context in which we will review Berkeley's discussion of "force" in *De Motu*. An example illustrates the issue. It would seem plausible, within the context of Newtonian Physics, to ask why the planets move in elliptical orbits rather than rectilinear paths in space. And the question is quite distinct from the question of what mathematical function relates the change in momentum of the planet to its mass, the mass of some other body, and the distance between them. The plausibility of the first question stems from the assumption that the elliptical motions are "un-natural," which means, in part, following Brian Ellis, that they require an explanation. "Forces," then, function to explain or account for such un-natural motions. For Ellis, this is the root of the distinction between a "kinematics" (a description of motion) and a "dynamics" (an account or explanation of motion in terms of forces).

When we speak of a law of motion, we do not speak of the way in which bodies actually move. Rather we speak of the way in which they would move, given that they are subject to the action of certain forces. Laws of motion, therefore, whether they by Newtonian, Aristotelian, or relativistic are laws of dynamics rather than kinematics.[51]

The problem, however, as we have indicated, is that certain "forces" cannot be said to refer to anything sensibly isolable from their effects. Such forces then appear to refer to something that is in principle unobservable. An overly-simplified view of Berkeley's theory of meaning would find this reason enough for Berkeley to deny the existence of "forces." Yet such a view would, among other limitations, fail to distinguish "occult" or "immanent" animating "causes" of motion (which Newton too rejects) from "impressed" forces such as "impact," "gravity" or "magnetism." Berkeley himself, as we will see, is not clear about the distinction.

[51] Ellis, *op. cit.*, p. 29.

That Berkeley is concerned with more than the problem of "unobservability" is indicated in the following passage from *De Motu*:

Again, force, gravity, and terms of that sort are more often used in the concrete (and rightly so) so as to connote the body in motion, the effort of resisting etc. But when they are used by philosophers to signify certain natures carved out and abstracted from all these things, natures which are not objects of sense, nor can be grasped by any force of intellect, nor pictured by the imagination, then indeed they breed errors and confusion. (*De Motu* 6)

Unfortunately Berkeley lumps together in this passage "force" in general and "gravity," which is an "impressed" force. However, it is significant that Berkeley contends that "force" as a specific "nature" is not only not observable, (not an "idea") but that it "cannot be grasped by any force of intellect," suggesting that the concept is not even intelligible as an explanatory principle.

It becomes clear that Berkeley's critique of the concept of "force" in the early sections of *De Motu* concerns the question whether material bodies can contain within themselves the principles (agency) of their own motion. "Force" used here connotes an immanent and dynamic principle, corresponding to the Aristotelian substantial form; and in fact, what Berkeley, following Newton, styles an "occult quality." [52]

---

[52] Koyre, *op. cit.*, (Appendix B pp. 139-148) documents the various senses the term "occult" had in the debates and polemics surrounding the Newtonian theory. Defenders of the Newtonian system like Cotes considered that it was the hypotheses of the Cartesian vortices that introduced "occult" qualities. Yet such "mechanical" hypotheses, as speculative and *ad-hoc* as they were, had empirical consequences, and Newton's substantial critique of vortex theory to account for central forces (*Principia* – third book) demonstrated that the consequences of the theory were not compatible with Kepeler's laws. Leibniz's critique of Newton appears to equate "occult" with "miraculous." (Koyre, 139-140) Here the issue is "action at a distance." Leibniz accepts the principle of inertia as describing "force-free" motion, and contends some mechanism must account for the planets being held in their orbits; the only allowable mechanism being some sort of impact phenomena. The point is emphasized by Leibniz in his correspondence with Newton's English defender Samuel Clarke. (See *Leibniz Clarke Correspondence,* ed. H. G. Alexander, (Manchester: Manchester University Press; 1956) Leibniz's third letter (sec. 16) and fourth letter (sec. 45).) Clarke makes a distinction between "natural" or law-like (regular) motions which are not reducible (to be explained in terms of) to the laws of impact, and motions which can be explained in terms of impact. Leibniz rejects this distinction; without a mechanism, the planetary orbits would require the continual intervention of God, hence the motions would be miraculous. For example, Leibniz's fourth letter, (sec. 45) "Tis also a supernatural thing, that bodies should attract one another at a distance, without any intermediate means; that that a body should move round, without receding in the tangent, though nothing hinders it from so receding. For these effects cannot be explained by the nature of things."

Newton himself, in the *Principia* (*op. cit.*, p. 371) distinguishes between "metaphysical" and "physical" hypotheses; or between "occult" and "mechanical" qualities. (All such hypotheses are to be rejected.) The clear suggestion is that the vortex hypothesis, although to be rejected is not considered making use of an "occult"

Berkeley's critique of endowing matter with a "vital" principle, or immament and efficient cause of motion, is both metaphysical and methodological. The former refers to his view that the properties which define "body" – extension, shape, solidity, motion or rest – are "ideas" and therefore passive. An immament moving force is only properly applied to "spirit." Historically this places Berkeley, not in disagreement, but in broad agreement with his corpuscularian opponents, whether Cartesian or Newtonian, who also would deny that the ultimate particles which constitute the material universe are moved by anything other than "impressed" forces (with the significant exception, perhaps, of "inertial" motion). This metaphysical objection, however, is rather weak; Berkeley allows that certain collections of "ideas," for example, the human body, are animated by a will other than the divine will. The question whether other collections of "ideas" which we style particular macroscopic (or "sensible") bodies are so animated would seem to be an empirical question; a negative answer would not seem to be required by Berkeley's metaphysical position. If for an Aristotelian the turning of an acorn into an oak is evidence of a vital and formative (and unobservable) principle of change, can we rule this out under the principles of immaterialism if we do not rule out that certain movements of the limbs are evidence of an animating will?

With respect to methodological issues, "occult causes" were considered useless by Berkeley, as by Newton, because they were believed to have no explanatory value. In this respect Berkeley follows the general critique of Aristotelian concepts (with the significant exception of Leibniz) in the seventeenth and eighteenth centuries. Discussing "gravity," Berkeley says:

obviously then it is idle to lay down gravity as the principle of motion; for how could that principle be known more clearly by being styled an occult quality? What is itself occult explains nothing. (*De Motu* 6)

quality. In "Query 31" of the *Optics* (*op. cit.*) there is a clear identification of "occult quality" with the Aristotelian "substantial form." A distinction is made between "gravity" which he styles a "manifest" quality and hidden or "occult" qualities. "And the Aristotelians gave the name of occult qualities only as they are supposed to lie hid in bodies, and to be the unknown causes of manifest effects. Such as would be the causes of gravity, and of magnetic and electric attractions, and of fermentations, if we should suppose that these forces or actions arose from qualities unknown to us, and incapable of being discovered and made manifest. Such occult qualities put a stop to the improvement of natural philosophy, and therefore of later years have been rejected. To tell us that every species of things is endowed with an occult specific quality by which it acts and produces manifest effects; is to tell us nothing; but to derive two or three general principles of motion from phenomena, and afterwards to tell us how the properties and actions of all corporeal things follow from those manifest principles, would be a very great step in philosophy, though, the causes of these principles were not yet discovered." It is this sense of the term "occult" that is accepted and used by Berkeley.

On this basis, Berkeley castigates Leibniz for introducing what the considers to be a vital principle within matter itself.

Leibniz likewise in explaining the nature of force has thus 'Active primitive force' which corresponds to the soul or substantial form. (*De Motu* 8)

Or:

Likewise Leibniz contends that effort exist everywhere, and always in matter, and that it is understood by reason where it is not evident to the senses. (*De Motu* 19)

A comparison in any detail between Berkeley and Leibniz is beyond the scope of this essay. We would note, however, in anticipation of later discussions, that Leibniz's attempt to combine a relativistic notion of the measure of motion with his acceptance of the reality of "force" as a principle of motion (or repetition in space) is quite close to Berkeley's position in the latter part of *De Motu*.[53]

Inertia, Impact, and Gravity

In section 23 of *De Motu* Berkeley comments:

And so about body we can boldly state as established fact that it is not the principle of motion. But if anyone maintains that the term body covers in

[53] Leibniz's arguments for an "active force" in matter can perhaps be viewed from either a physical or mathematical perspective. With respect to the first, as Jammer points out, Leibniz accepted that although in determining motion, the choice of reference frame was a matter of convention, real motion could be distinguished from apparent motion, in that the former referred to a body subject to a force. (Jammer, *Concepts of Force, op. cit.*, pp. 161-163) For Leibniz, neither the inertial properties of bodies nor their impenatrability can be explained in terms of extension alone. Mere change of place, then is not sufficient to explain real as opposed to apparent motion. From a mathematical perspective, Leibniz's conception of force can be viewed as suggesting an answer to the problem of the "continuum," a problem that occupied Berkeley as well. If we attempt to understand a finite motion as the successive occupation of an infinite number of points, it appears impossible to achieve a finite distance, since the measure of each point is zero. Leibniz essentially argues that the moving body must be considered to be passing through each point. At every instant then the body must be said to be moving, so that at each instant the body is endowed with an "active force," or "nisus for change." ("Specimen Dynamicum pro admirandis Naturae Legibus circa corporum vires et mutuas act ones detergendis et ad suas causas revocandis," Pars I) (*Acta Eruditorum* Lips an. 1695) English translation by Alfred G. Langly), printed in Phillip P. Wiener, ed. *Leibniz, Selections,* (Charles Scribner's and Sons, 1951) pp. 119-137. "Thus strictly speaking, motion just like time, when reduced by analysis to its elements, has no co-existence as a whole so long as it possesses no co-existing parts. And thus there is nothing real in motion itself apart from the reality of the momentary transition which is determined by means of force and a nisus for change..." Berkeley will reject in dynamics the concept of an "impetus" as he will reject as unintelligible the concept of an instantaneous velocity. His own solution of the "paradoxes" of the continuum will be to deny that any finite segment of extension or motion can be infinitely subdivided.

its meaning occult quality, virtue, form or essence besides solid extension and its modes, we must leave him to his useless disputation with no ideas behind it, and to his abuse of names which express nothing distinctly.

The fact that bodies move is not sufficient to establish an internal animating "force" or principle of motion. The passivity of body extends to body in motion as well as body at rest. Berkeley appears to buttress his view that body in motion is not dynamically distinct from body at rest, by appeal to Newton's first law of motion, or the principle of inertia.[54]

Body in fact persists equally in either state, whether of motion or of rest. Its existence is not called its action; nor should its persistence be called its action. Persistence is only continuance in the same way of existing which cannot properly be called action. (*De Motu* 27)
The most learned philosophers of this age lay it down for an indubitable principle that every body persists in its own state, whether rest or uniform motion in a straight line, except in so far as it is compelled from without to change that state. (sec. 33 *De Motu*)
Modern thinkers consider motion and rest in bodies as two states of existence in either of which every body, without pressure from external force would remain passive. (*De Motu* 34)

The principle of inertia, however, not only does not rule out efficient causality within the phenomal world; it might, from a certain point of view, be said to require it. For example, we designate a certain class of motions as "inertial," understanding by this that the body moving is not acting under any impressed forces. Such motions, following Brian Ellis, can be called natural motions.[55] A system is in an un-natural state if its being in that state requires causal explanation and a system is said to require causal explanation if subsuming its states under a law of succession is not considered sufficient to explain it. A "law of succession" would be some law that allows us to predict future states of the system from knowledge of its present state. Both uniform velocity and uniform acceleration can be subsumed under a law of succession, but the latter is not considered to be sufficiently explained by such a law. The uniform acceleration of free-fall can be represented by a mathematical function relating distance and time, but such a law could not be said to explain "why" a body dropped near the earth accelerates.

"Forces," then, have the conceptual sense of that which is introduced to "explain" un-natural motions (or states, like the deformation of ma-

---

[54] Newton, *Principia, op. cit.,* Book I, "Axioms or Laws of Motion." Law I: "Every body continues in its state of rest, or of uniform motion in a right line, unless it is compelled to change that state by forces impressed upon it."
[55] Ellis, *op. cit.,* p. 45.

terials – the water, for example, in Newton's whirling bucket). This is a distinct issue from the question of the psychological genesis of our concept of "force," which may very well be the experienced resistance of objects to motion; and it is a distinct issue from the question of how "forces" are measured.

It turns out that the principle of inertia is fundamentally irrelevant to Berkeley's case against immanent efficient causes ("forces") of motion, and his use of the principle is either merely *ad hominem* or misguided. His point is that uniform motions, whether uniform velocity *or* acceleration, offer no evidence for a vital principle in matter, understanding by "vital principle" a "principle or faculty" for instituting or stopping motion, within the moving body itself.

Heavy bodies are born downwards, although they are not affected by any apparent impulse; but we must not think on that account that the principle of motion is contained in them ... All heavy things by one and the same certain and constant law seek the center of the earth, and we do not observe in them a principle or any faculty of halting that motion, of diminishing it or increasing it except in fixed proportion, or finally of altering it in any way. They behave quite passively. Again in strict and accurate speech, the same must be said of percussive bodies ... (*De Motu* 26)

Here, however, the motion being discussed is uniform acceleration, not uniform velocity. Since the principle of inertia singles out a quite restricted class of "lawful" motions as force-free, those described as uniform velocity, it has no particular relevance to the question of whether we can impute vital forces to bodies undergoing uniform acceleration. And observations of a textual nature offer some evidence that Berkeley associates the *passivity* of "matter" in motion more with the "*lawfulness*" of the motions studied by astronomy and mechanics than with anything distinctively entailed by the principle of inertia. We mentioned *De Motu* 34, where Berkeley uses the phrase, "modern thinkers consider motion and rest, ..." and does not use the phrase, "modern thinkers consider *uniform* motion and rest." The same failure to distinguish uniform velocity and acceleration occurs in *De Motu* 51:

For since experience shows that it is a primary law of nature that a body persists exactly in 'a state of motion and rest as long as nothing happens from elsewhere to change that state,' and on that account it is inferred that the force of inertia is under different aspects either resistance or impetus,[56] in this sense assuredly a body can be called indifferent in its own nature to motion or rest. Of course it is as difficult to induce rest in a moving body as motion in a resting body; but since the body *conserves* equally either state, why should it not be said to be indifferent to both? [Our en phasis]

What is "conserved," however, is not any "lawful" motion (motion that can be subsumed in Ellis's phrase under "a law of succession,") but rather (under the principle of inertia) rest or uniform velocity in a straight line. If the phrase "primary law of nature" refers to the principle of inertia, the law itself, without some added assumptions about the source of the "impressed forces," cannot tell us anything about whether bodies undergoing accelerated motion are animated. A body undergoing uniform acceleration is undergoing a lawful motion. If Berkeley is contending that a body will "conserve" any "lawful motion" unless acted upon from "elsewhere," then he clearly misunderstands the principle of inertia.

It is unlikely, however, that Berkeley, who knew the *Principia* well, misunderstood the principle of inertia. More likely, he understood it to be making two distinct points, neither of which entailed the other. The first is that bodies undergoing uniform velocity in a straight line need not have their motions explained in terms of any "forces"; and the second point is that bodies undergoing accelerations must have their motions explained in terms of "impressed" or "external" forces. "External" has the sense of from without the body moved, as opposed to "immament," or a force from within the body. However, the term "impressed" does not necessarily entail "external" in this sense, since even Berkeley would allow that the will impresses a change in motion on the limbs. To associate "impressed" with "external" is to accept the general Newtonian (and Cartesian) view that "body" or "matter" is passive; that it cannot in any way comparable to an act of "will" institute a change in its own motion. Another way of expressing this is to suggest that it is the conception of the passivity of matter which entails a particular interpretation of "impressed" in the first law of motion, rather than that law, in any significant sense, implying the passivity of matter.

Since Berkeley accepts the principle of inertia, there remains the question of how he understands the nature of "impressed forces" (forces that would be introduced to explain un-natural motions). How are phrases such as "compelled from without" (*De Motu* 33), or "without pressure from external force" (*De Motu* 34) to be interpreted?

From the point of view of what we have termed "strict mechanism," changes in the natural motion of a body were attributed to impacts of collisions with other material bodies. Yet to give such "impacts" the status of efficient causes, would appear to violate Berkeley's general principle that efficient causality can only be attributed to will.

---

[56] See Newton, *Principia, op. cit.,* Definition III (p. 5).

It remains to discuss the cause of the communication of motions. Most people think that the force impressed on the moveable body is the cause of the motion in it. . . . It is clear, moreover that force is not a thing certain and determinate, from the fact that great men advance very different opinions, even contrary opinions about it, and yet in their results attain the truth. For Newton says that impressed force consists in action alone, and is the action exerted on the body to change its state, and does not remain after the action. Torricelli contends that a certain heap or aggregate of forces impressed by percussion is received into the mobile body, and there remains and constitutes impetus. . . . But although Newton and Torricelli seem to be disagreeing with one another, they each advance consistent views, and the thing is sufficiently well explained by both. For all forces attributed to bodies are mathematical hypotheses, just as are attractive forces in planets or sun. But mathematical entities have no stable essence in the nature of things; and they depend on the notion of the definer. Whence the same thing can be explained in different ways. (*De Motu* 67)
And so it comes to the same thing whether we say that motion passes from the striker to the struck, or that motion is generated do novo in the struck and is destroyed in the striker. (*De Motu* 68)

These passages are worth some detailed comment with respect to a number of issues that they raise. In the first place, the laws of impact do not, for Berkeley, have a privileged place as causal accounts for the changes in the momentum of a body, as they would have for Descartes and Leibniz. If we deny efficient causality to collisions, there is no longer the pressure to reduce all explanations for the changes in momentum of a body to the laws of impact. Passages previously mentioned from the *Siris* document this contention that Berkeley would allow *other* fundamental principles in mechanics than the laws of impact (for example, principles of attraction and repulsion). Berkeley, unlike Leibniz, had to particular problem with "action at a distance."

In addition the passages suggest that the concept of "force" can be completely dispensed with in mechanics. There are certain "mechanical" laws, for example Newton's third law of motion, where, although force terms are used, they could be dispensed with. Thus in a two-body collision, one can state that after impact the change in momentum of one body will be equal in magnitude and oppositely directed to the change in the momentum of the other body. Force terms like "action" and "reaction" in the Newtonian expression of the third law of motion are elliptical, and express nothing of real (efficient) causation.

I, for my part, will content myself with hinting that that principle [Newton's third law of motion] could have been set forth in another way. For if the true nature of things, rather than abstract mathematics, be regarded, it will seem more correct to say that in attraction or percussion, the passion of bodies, rather than their action, is equal on both sides. For example, the stone tied by a rope to a horse is dragged towards the horse just as much as the horse towards the stone; for the body in motion impinging on a quiescent body suffers the same change as the quiescent body. And as regards real effect, the striker is just as the struck, and the struck as the striker. And that change on both sides, both in the body of the horse, both in the moved and in the resting, is mere passivity. It is not established that there is force, virtue or bodily action truly and properly causing such effects. The body in motion impinges on the quiescent body; we speak in terms of action and say that that impels this; and it is correct to do so in mechanics where mathematical ideas, rather than the true natures of things, are regarded. (*De Motu* 70) [57]

To attribute "force" or efficient causal action to the "moving body" in collision phenomena (where presumably one body is at rest, or like the "stone" in Newton's example, is being "dragged") is not incorrect, Berkeley suggests, if "force" means no more than a "mathematical hypothesis"; just as it is correct to attribute "attractive" forces to the planets or the sun, if we mean no more by this than a "mathematical hypothesis." (*De Motu* 67) However, it is not clear what Berkeley is referring to by a "mathematical hypothesis" either with respect to impact or gravitational phenomena. One possibility is that he means a particular mathematically formulated law, for example, the law of gravitation, relating the change in the momentum of a body to its mass, the mass of some other body and the distance between them; or the laws of impact which can also be formulated without the use of dynamic, or force terms like "action" or "reaction." Although Berkeley will often argue that the purpose of physics is to formulate such laws, which then function as premises for the "demonstration" of phenomena,[58] it seems unlikely that he makes synonymous, for example, gravitational attraction and the law of gravitation, or "action and reaction," with Newton's third law of motion. The laws of motion, then, are not themselves what Berkeley means by "mathematical hypotheses."

Since Berkeley's discussion of mechanics so often parallels that of Newton (even considering their differences), we are afforded an insight into what Berkeley means by "mathematical hypothesis," by showing how Newton would have used this, or some comparable expression. Berkeley indeed makes reference to Newton early in *De Motu*.

As for attraction, it was certainly introduced by Newton, not as a true physical quality, but only as a mathematical hypothesis. (*De Motu* 17)

---

[57] Newton, *Principia*, Law III (p. 14).
[58] See for example, *De Motu* (71).

Berkeley likely had in mind definition VIII, Book I of the *Principia*.[59]

I likewise call attraction and impulse, in the same sense, accelerative and motive; and use the words attraction, impulse, or propensity of any sort towards a center, promiscuously and indifferently, one for another; considering these forces not physically, but mathematically; wherefore the reader is not to imagine that by these words I anywhere take upon me to define the kind or manner of any action, the causes or the physical reasons thereof, or that I attribute forces in a true and physical sense, to certain centers (which are only mathematical points) when at any time I happen to speak of centers as attracting, or as endowed with active powers.

What Newton means is that it is useful from a mathematical point of view to consider "point masses" as exercising some "attractive" power which draws other masses from their rectilinear (normal) path. The law of gravity does not require, however, as regards the *real* nature of this "force," any physical hypothesis. By a "physical hypothesis" Newton is undoubtedly referring to vortex theories and other "mechanisms" which attempt to "explain" central forces without positing action at a distance.

Although Newton in the *Principia* offers no "mechanical" hypothesis about how the force of gravity (or what Koyre perhaps more appropriately labels a "vis centripeta") operates,[60] he never claims that this "force" is a mere "mathematical hypothesis," or heuristic fiction, as one might claim in the case of the "lines and angles" of geometers, or even in the case of the "insensible particles" of corpuscular hypotheses.[61] And clearly, given his view of the principle of inertia, he could not make (contra Berkeley) such a claim. If a body acted on by no "impressed" forces moves at a uniform velocity in a straight line, then it follows logically that bodies whose paths are closed conics, are acted upon by impressed forces. Moreover, as modern commentators have pointed out, Newton understood his second law to be an empirical hypothesis and not a definition of "impressed force." [62] This suggests that for him "forces" were physically real and certainly conceptually, if not sensibly distinct from their effects. That one could not sensibly identify a force independently of its effects (this would not be true, however, of "kicks, thrusts, pushes, and impacts of other varieties which could perhaps be considered sensibly distinct from

[59] Newton, *Principia*, *op. cit.*, p. 9.

[60] Koyre, *op. cit.*, p. 153.

[61] See for example, Newton, *Principia*, "Scholium" to Definition VIII, *op. cit.*, p. 11. "The causes by which true and relative motions are distinguished one from the other, are the forces impressed upon bodies to generate motion."

[62] Arguments that Newton's "Laws" (axioms) of motion were meant to be empirical generalizations can be found, for example, in Ellis, *op. cit.*, and K. A. Koslow, "The Law of Inertia: Some Remarks on its Structure and Significance," in *Philosophy Science and Method*, ed., Morgen Besser, Suppes, White, *op. cit.*, pp. 549-568.

their effects) [63] would not for Newton be a mark against its reality. It is a mistake to equate, then, Newton's positivist caution concerning the modus operandi of "forces" like "gravitation" or other central forces, with a view that "forces" are mathematical fictions (as are idealizations like "point masses").

We will return shortly to the sections in *De Motu* concerning "gravity." With respect, however, to impact phenomena, (and the laws of impact) Berkeley's suggestion in *De Motu* (section 68) that alternative hypotheses which appeal to forces are equivalent, if they entail the same kinematic law, is worthy of some comment. Berkeley would be correct (ignoring now issues concerning simplicity of hypotheses etc.) if, in fact, there was no other empirical consequence of the hypotheses, or if all the empirical consequences were the same for either hypothesis. For example, with respect to collisions, is there any observation compatible with "the motion of (A) is 'communicated' to (B)," and not logically compatible with, "the motion of (A) 'ceases' at impact, and begins '*de novo*' in (B)?" If, for example, (B) were deformed in some way and the amount of deformation was proportional to its change in momentum after impact, would this be an argument for the "communication of motion?" There is, interestingly enough, a passage in *De Motu* where Berkeley appears to give an affirmative answer:

We feel at times the pressure of a gravitating body. But that unpleasant sensation arises from the motion of a heavy body communicated to the fibres and nerves of our body and changing their situation; and therefore it ought to be referred to percussion. (*De Motu* 13)

Berkeley could of course argue that we are being overly critical, and that his use of "communicate" here is elliptical. What we observe are correlations between changes in momentum of some body, deformations in the nervous system, and certain sensations. This is compatible with Berkeley's view that aside from our reflective awareness of the effort of will, we have no intuition or sensible observation of "force" or "power." In the same sense there can be no observation of the *"communication"* of motion sensibly distinct from the motions of the bodies themselves. Berkeley's claim that, in effect, there is no empirical signification to the expression "communcation of motion" is a consequence of his view that there is no efficient causality in the natural world; at least with respect to matter. What is historically important here is that the metaphysics of

---

[63] On the question of whether the existence of forces logically entails the existence of force effects, see Ellis, *op. cit.,* and I. E. Hunt and W. A. Suchting, "Force and Natural Motion," *op. cit.*

immaterialism has again liberated Berkeley from accepting what had become almost an axiom among his immediate predecessors and contemporaries in physics and philosophy, that changes in motion could only be effected through impact.

Unlike impacts, as we have mentioned, central forces like "gravity," "magnetism," etc., cannot be even sensibly distinguished from their alleged "effects." As a consequence the following remarks in *De Motu* about the role of the "physicist" could only apply to the study of impact phenomena:

> The physicist studies the series or successions of sensible things, noting by what laws they are connected, and in what order, what precedes as cause, and what follows as effect. And on this method we say that the body in motion is the cause of the motion in the other, and impresses motion on it, draws it also or impels it. (*De Motu* 17)

Since there is nothing in gravitational phenomena which would function as a sensible referent for what is elliptically called a "force" (like an impact or collision) and which would be temporally antecedent to its effects, the concept of "causality," as Berkeley wishes to use it, could not apply to such phenomena. There is no distinct sensible mark of such "forces," so that we cannot speak of "causality" here in the sense of invariably correlated but temporally distinct phenomena.[64]

In addition, with respect at least to planetary motions, we have no way other than observation of changes in the momenta of the planets to demonstrate that they are in a gravitational field.[65] Noting this, Berkeley contends that

> The force of gravitation is not to be separated from momentum; but there is no momentum without velocity, since it is mass multiplied by velocity; again velocity cannot be understood without motion, and the same holds therefore of the force of gravitation. Then no force makes itself known except through action, and through action it is measured; but we are not able to separate the action of a body from its motion; therefore as long as a heavy body changes the shape of a piece of lead put under it, or of a cord, so long it is moved; but when it is at rest, it does nothing, or (which is the same thing) it is prevented from acting. In brief, those terms *dead force* and *gravitation* by the aid of metaphysical abstraction are supposed to mean something different from moving, moved, motion, and rest, but in point of

---

[64] In "free-fall" we can speak of antecedent and consequent states connected by a law, but it seems doubtful that Berkeley would apply the term "causal" to such laws.

[65] One might contend that theoretically there are other ways, although practically impossible. For example, Nagel remarks: "It is fanciful imagination, not physics which would measure the imputed force acting on a planet by means of an extended spring connecting the planet and the sun." Nagel, *Structure of Science, op. cit.,* p. 189.

fact, the supposed difference in meaning amounts to nothing at all. (*De Motu* 11)

Jammer quotes the passage in support of his claim that Berkeley views the study of motion merely as a kinematics, identifying the "force" of gravity with the "law" of gravity.[66] However, this interpretation ignores the somewhat equivocal character of section 11, and the previous section where Berkeley comments:

We must, however, admit that no force is immediately felt by itself, nor known or measured otherwise than by its effect; but of a dead force or of simple gravitation in a body at rest, no change is taking place, there is no effect. (*De Motu* 10)

Both sections 10 and 11 might be interpreted as allowing the existence of "forces," although contending that such forces can only be known (or "measured") by means of their "effects." Such a view might allow that central forces like gravitational and magnetic force are dispositional in character; that, for example, although the magnet is continually exercising its "force," the latter is only made manifest when other bodies (iron filings, for example) are placed near it. Yet it appears that it is precisely this "dispositional" character (which might imply the existence of "forces" distinct from "effects") that is denied by Berkeley. By "dead force" he probably means what we now call "potential energy," the "energy" of a stone, for example, that has been raised to a certain height. Modern physics has given a precise operational significance ot the term[67] that undoubtedly would have been satisfactory to Berkeley; his point being that it is meaningless to impute any "force" or "power" to the stone when it is motionless. The point, however, is still ambiguous; it is still not clear whether, when the stone is moving we can impute "force" to it as the cause of certain manifest effects, or whether "force" means no more than the set of such "effects." A comparable point can be made about the issue of whether a body at rest, for example, with respect to the center of the earth, can be said to be "attracted" by that center. If we understand the passages correctly, Berkeley would find the claim of gravitational "attraction" on such a body to be meaningless.[68] However, it is not clear

[66] Jammer, *The Concept of Force, op. cit.*, p. 205.
[67] In general "energy" can be defined as the "capacity" for doing "work," and "work" is defined as force multiplied by distance moved at the point of application of the force in the direction of the force. "Potential" energy can be thought of, for example, as the capacity a body has for doing work by virtue of its position (i.e., gravitational "potential energy" a body has by virtue of its position above the ground).
[68] It is not always clear whether Berkeley is saying that the existence of "forces" logically entails the existence of force – effects, where some conceptual distinction is still made between forces and force-effects; or is saying that the term "force" means

from the passages (*De Motu* 10 and 11) whether gravitational "attraction" *means* nothing more than the class of its "effects."

There are passages in *De Motu* which suggest that Berkeley does take this latter position.

As for gravity, we have already shown above that by that term is meant nothing we know, nothing other than the sensible effect, the cause of which we seek. (*De Motu* 22)

The difficulty is understanding what Berkeley means by the term "cause" in the passage. Clearly, "efficient cause" is not meant. More plausibly, Berkeley is echoing Newton who distinguishes gravity as a "manifest quality," (what we would call gravitational phenomena, the tides, planetary orbits, free fall etc.), and the "cause" of this quality.[69] "Cause" here would have the sense of modus operandi; what "mechanism" (impact phenomena, vortices etc.) can explain the un-natural motion, for example, of the planets? Alternatively expressed, the search for a "cause" might be considered the attempt to show that the *law* of gravity is deducible from allegedly more encompassing laws; for example, the laws of impact. If our interpretation is correct, by the time of the *Siris*, Berkeley has evidently come to acknowledge that principles (or laws) of attraction and repulsion (those that apparently deal with action at a distance) may be as fundamental as the Cartesian laws of impact.[70]

In section 38 of *De Motu* we have perhaps a clearer statement that "force" terms refer to nothing in reality:

And just as geometers for the sake of their art make use of many devices which they themselves cannot describe nor find in the nature of things, even so the mechanician makes use of certain abstract and general terms, imagining in bodies force, action, attraction, solicitation, etc. which are of first utility for theories and formulations, as also for computations about motion, even if in the truth of things, and in bodies actually existing, they would be looked for in vain, just as the geometers' fictions made by mathematical abstraction.

However, as we suggested in our discussion of *De Motu* (18), Berkeley never makes it clear how the concept of gravitation as a "force" functions as an aid (or heuristic device) in the formulation of the law of gravity. If (as in section 18) Newton is being referred to as the "mechanician"; the heuristic device was not the introduction of the concept of "force,"

no more than what we have previously called force effects (e.g., the change in momentum of a "gravitating body").

[69] Newton, *Optics, op. cit.,* "Query 31."
[70] *Siris* (162, 245).

but rather the idealization of considering this "force" to be "attractive" rather than "impulse" and located at the geometric "centers" of the masses involved. Newton had fundamentally a four-fold distinction: (a) the phenomena of gravitation (gravity as a "manifest quality"), (b) the "force" which "caused" (a), (c) the mode or mechanism by which this "force" operated, and (d) the law of gravity; that is the "force function" (using the terminology of Ernest Nagel) [71] which showed the change in momentum of the planets to be a mathematical function of its mass, the mass of the sun, and the distance between them. Newton never denied (b), the existence of a "force," since all non-inertial motions required the existence of impressed forces.[72] The law of gravity relates the "effect" of this force (change in momentum of a planet, for example) to certain other parameters (Masses, distances, etc.)

The above remarks are relevant in evaluating Berkeley's contention in the *Principles* (sec. 103-105) that the law of gravitation is an inductive generalization. He suggests that a "philosopher" (Newton?)

having observed a certain similitude of appearances, as well in the heavens as the earth, that argue innumerable bodies to have a mutual tendency towards each other which he denotes by the general name "attraction." (*Principles* 104)

However, with respect to the relation of the planets to the sun, we do not observe this "mutual tendency" of bodies towards one another. It was the genius of Newton to demonstrate that the motion of free-fall and the orbital motions of the planets were encompassed by the same law.[73] That we find this "largeness of comprehension" in Newton and not in Galileo cannot be grounded in historical accident. For Galileo, inertial motion (that which requires no explanation in terms of impressed forces) was circular; consequently the orbits of the planets would not be the kind of phenomena that would suggest the existence of such forces.

The comprehensive vision of Newton, then, cannot be separated from his conception of inertial motion, formulated in his first law. The importance of conceptual underpinnings in determining even what questions one asks of nature, or how one "sees" phenomena, has been well documented, and we will not discuss it in detail.[74] What is important here is

---

[71] Nagel, *op. cit.*, pp. 189-192.

[72] See for example, Newton, *Optics, op. cit.*, "Query 31," p. 76. "What I call Attraction may be perform'd by impulse, or by some other means unknown to me. I use that Word here to signify only in general any Force by which bodies tend towards one another, whatsoever be the Cause."

[73] Newton, *Principia, op. cit.*, Definition V, p. 6.

[74] For example; Thomas Kuhn, *The Structure of Scientific Revolutions*, (Chicago: University of Chicago Press, 1962).

that the Newtonian system (certainly as interpreted by Newton) is based on a distinction between "force-free" and "force" – determined motion. It is Newton's understanding of the principle of inertia which entails that impressed forces are required to explain the planetary orbits.

Berkeley at times, however, confuses the question of the attribution of forces with the question of accounting for the operation of forces. Elsewhere in the *Principles,* when discussing gravitational phenomena (the mutual drawing together of bodies) he remarks:

But how are we enlightened by being told this is done by attraction? Is it that that word signifies the manner of the tendency; and that is by the mutual drawing of bodies instead of their being impelled or protruded towards each other? But nothing is determined of the manner or action, and it may as truly (for ought we know) be termed "impulse," or "protrusion," as "attraction." (*Principles* 103)

Here Berkeley appears to be echoing Newton's caution in the *Principia* as to the mechanism by which the bodies are directed out of their normal paths. Newton himself hesitates to endow bodies with an "attractive" power that can act across empty space. He never, however, identified this "vis centripeta" with its "effects," (whatever the mode of its operation). As illustration, one can look to the discussion of "centripetal force" in Book I, Definition V of the *Principia.* We quote one passage:

And after the same manner that a projectile by the force of gravity, may be made to resolve in an orbit, and go round the whole earth, the moon also, either by force of gravity, if it is endued with gravity, or by any other force, that impels it towards the earth, may be continually drawn aside towards the earth, out of the rectilinear way which by its innate force it would pursue; and would be made to revolve in the orbit which it now describes; nor could the moon without some such force be retained in its orbit.[75]

Berkeley, then, although he appeals to the principle of inertia as evidence for the "passivity" of matter in motion, evidently does not recognize that the "principle," as formulated by Newton, requires the attribution of "forces" to account for non-uniform motion. Or, if Berkeley recognizes the need for the attribution of "impressed forces" to explain certain motions, he distinguishes such "forces" from "vital forces," that is, immanent principles of movement. However, it is not clear that this distinction (one with which Newton would agree) is one of substance *in the context* of Berkeley's general aim to eliminate the concept of "efficient cause" from the scientific study of nature. How are we to understand

Norwood Russell Hanson, *Patterns of Discovery,* Cambridge: Cambridge University Press, 1961.
[75] Newton, *Principia, op. cit.,* Definition V.

these "impressed forces" if not as efficient causes of, for example, the non-rectilinear motion of the planets? Certainly for Newton the "vis centripeta," termed "gravitational force" (as all "impressed forces") was not to be identified with its "effects," but was the cause of such effects.[76]

It is interesting, in this regard, to note that Berkeley appeared to accept Newton's "laws" ("axioms") of motion as empirical propositions (as evidently Newton did himself). In *De Motu*, for example, he distinguishes between a metaphysical and physical sense to the term "principle."

The true, efficient, and conserving cause of all things by supreme right is called their fount and principle. But the principles of experimental philosophy are properly to be called foundations and springs, not of their existence but of our knowledge of corporeal things ... Similarly in mechanical philosophy those are to be called principles, in which the whole discipline is grounded and contained, those primary laws of motion which have been proved by experiments, elaborated by reason and rendered universal. These laws of motion are conveniently called principles, since from them are derived both mechanical theorems and particular explanations of the phenomena. (*De Motu* 36)

If the first law of motion in the *Principia* is considered one of these "primary laws," then it is not (from Berkeley's point of view) simply an implicit definition of the absence of impressed forces, but rather the "generalization" of an experimentally determined "fact," that in the absence of "impressed forces" bodies maintain a uniform velocity in a straight line. If this is the case, however, it is difficult to see how Berkeley could consider the phrase "impressed forces" either as "meaning" the same thing as their "effects," or as elliptical for "mathematical hypothesis." Admittedly, again, "impressed forces" are not "vital forces." On the other hand, they appear to have the status of efficient causes; at least if one accepts Newton's first law as an empirical proposition, it is difficult to conceive what other status such forces would have.

Berkeley recognizes that to explain the "mutual tendency" of bodies towards one another, whether in the phenomena of gravitation or cohesion, by appeal to an "attractive" power that bodies have, is no explanation at all, and is akin to using "occult qualities" as explanatory principles (*Principles* 103). The physicist seeks to relate the "effects" of gravity (for example, the change in momentum of a planet) to other variables (masses, distances). With the law of gravity, as with the laws of

---

[76] *Ibid.*, it would difficult to draw any other conclusion from this discussion than that Newton considered "forces" to have the status of efficient causes.

impact, force terms ("attraction," "impuslion," etc.) can be dispensed with.

And just as by the application of geometrical theorems, the sizes of particular bodies are measured, so also by the application of the universal theorems of mechanics, the movements of any part of the mundane system, and the phenomena thereon depending, become known and are determined. And that is the sole mark at which the physicist must aim.

Of course aiming for a law that mathematically relates the effect of centripetal forces to masses and distances is compatible with the existence of centripetal forces; in fact, realizing that the same "force" is operating in qualitatively different domains is an impetus to unifying these domains under one law. It is true that establishing that a body is moving under impressed forces, in itself, gives no predictive power as to future states of the body. In this respect an "account" of motion in terms of a "law" of motion, must be distinguished from an "account" of motion in the sense of "explaining" why it is un-natural motion. "Acting under impressed forces," does explain *why* a body is accelerating rather than moving at uniform velocity in a straight line. This presupposes, again, that the expression "impressed force" has a semantic content [77] distinct from that which produces "un-natural motion." Such a presupposition takes Newton's second law, which relates impressed forces and changes in momentum, to be an empirical proposition. And, as we have indicated, this appears to be the way Berkeley takes the Newtonian laws of motion.

As we have mentioned, writers like Jammer believe Berkeley was headed in the direction of viewing physics merely as a kinematics (mathematical description of motion) where force terms would play no role. Jammer comments:

With the work of Mach, Kirchoff, and Hertz the logical development of the process of eliminating the concept of force from mechanics was completed. The development of mathematical physics from the time of Newton onward was essentially an attempt to explain physical phenomena in terms of their mass points and their spatial relations. Since the time of Keill and Berkeley, it became increasingly clear that the concept of force, if divested of all its extra scientific connotations, reveals itself as an empty scheme, a pure relational or mathematical function ... The concept of force in its metaphysical sense as causal transeunt activity had no place in the science of the empirically measurable.[78]

Although we would agree that Berkeley often appears to take something

---

[77] See also the discussion by Norwood Hanson, "Newton's First Law," in *Beyond the Edge of Certainty*, ed. Robert G. Colodny, *op. cit.*, pp. 6-29.
[78] Jammer, *Concept of Force, op. cit.*, p. 229.

like the above view, his discussion of "force" remains ultimately ambiguous, since he fails to clarify how he understands the Newtonian concept of "impressed forces," particularly as they function in the first and second laws of motion.

The above remarks lead naturally to an evaluation of the thesis that Berkeley is a precursor of Ernst Mach, not only with respect to the critique of the Newtonian concepts of absolute space, time, and motion, but with respect to Mach's critique of the concept of "force."

If Berkeley's general philosophic position, conjoined with the more specialized critique of mechanics found in *De Motu*, supports the view (as we believe it does) that the science of matter in motion, properly construed, is a kinematics, than the similarity to Mach is quite striking. In the *Science of Mechanics*, Mach comments:

> If we pass in review the period in which the development of dynamics fell – a period inaugurated by Galileo, continued by Huygens, and brought to a close by Newton – its main result will be found to be the perception that bodies mutually determine in each other accelerations dependent on definite spatial and material circumstance, and that there are masses ... In reality one great fact was established. Different pairs of bodies, determine, independently of each other, and mutually in themselves, pairs of accelerations, whose terms exhibit a constant ratio, the criterion and characteristic of each pair ... That which in the mechanics of the present day is called *force* is not a something that lies latent in the natural processes, but is a mesurable actual circumstance of motion. Also when we speak of the attractions or repulsions of bodies, it is not necessary to speak of any hidden causes of the motions produced.[79]

From Mach's point of view, then, there are only "laws" of motion; even material deformations ("distortion, compression of bodies")[80] cannot be said to imply the existence of "forces."

However, Mach clearly recognizes, as Berkeley did not, that the program of establishing mechanics as a kinematics requires a reformulation of the Newtonian first and second laws of motion. Particularly relevant is the reformulation of the principle of inertia. For Newton, the property of inertia – the power of resisting a change in motion – is an inherent property of all bodies, not relative to any other body in the universe. Newton could conceive in principle (if not operationally establish) a one-body universe; the motion of such a body would be characterized as "force-free" or "natural" motion. Mach's contribution is fundamentally to deny this ontological distinction between force-free

---

[79] Mach, *op. cit.*, pp. 306-307.
[80] *Ibid.*, p. 307.

and force-determined motion. We can, he suggests, know no more of
bodies than their motions (and the consequences of such motions) with
respect to other bodies.

In Mach's view, uniform or "inertial" motion does not have the distinction of expressing a type of motion unrelated to (not a consequence
of) the presence of other masses in the universe. A physical frame of
reference is required, for example the fixed stars, and we can think of
inertial motions as a function of that physical frame.

Instead of saying the direction and velocity of a mass U in space remain
constant, we may also employ the expression, the mean acceleration of the
mass U with respect to the masses, m, m′, m″. . . at the distances r, r′, r″. . .
is $= 0$, or $d^2 (\Sigma mr/\Sigma m)/dt^2 = 0$.[81]

This is, for Mach, an empirical proposition, but unlike Newton's formulation, does not refer to a property a single body would allegedly have in
a universe devoid of other masses. Inertial effects, like gravitational
effects, are conditioned by the spatial relations of masses; the laws of
mechanics are mathematical functions relating masses and distance, and
mechanical explanation consists in showing how phenomena are in conformity with these laws.

In the *Siris* which reflects the queries in Newton's *Optics,* we get,
perhaps more than in *De Motu,* something comparable to the Machian
view, that what we are dealing with are the motions of "masses" (motions
that are labeled "attraction" or "repulsion") and "laws" which relate
states of the moving body (e.g., change in momentum) to its distance
from other masses. To quote again the *Siris*:

The minute corpuscles are impelled and directed, that is to say moved to
and fro from each other, according to various rules or laws of motion. The
laws of gravity, magnetism, and electricity are diverse. (*Siris* 235)

Berkeley, however, (again reflecting the speculations in the "queries" in
Newton's *Optics*) posits an aether or "first corporeal principle," suggesting that as with Newton, "action at a distance" is not totally acceptable. With Mach there *is* the clearest emancipation from the view that the
changes in momentum of a body with respect to an "inertial frame of
reference" must be explained in terms of contact with another material
body. Again Mach:

Like all his prodecessors and successors, Newton felt the need of explaining
gravitation, by some such means as actions of contact. Yet the great success
which Newton achieved in Astronomy with forces acting at a distance as

[81] *Ibid.,* p. 287.

the basis of deduction, soon changed the situation very considerably. Inquires accustomed themselves to these forces as points of departure for their explanations and the impulse to inquire after their origin soon disappeared almost completely.[82]

Giving up the notion of "forces" as real entities, distinct from and causally related to their "effects," was even more radical. Although there are many passages pointing to this achievement in Berkeley, it is not a consistent achievement. We will see, in fact, when discussing the parts of *De Motu* which are devoted to the concepts of absolute space and motion, that Berkeley appears to introduce, in a significant way, the concept of "force" as efficient cause.

C. ABSOLUTE SPACE AND MOTION *(DE MOTU)*, *(PRINCIPLES)* [83]

It is helpful to distinguish in Berkeley's writings three different but related aspects of his analysis of the concept of "absolute space." The first we can term the problem of the meaning of the concept itself; the second concerns the question of what is the object of geometry; and the third concerns Berkeley's critique of Newton's attempt to find an empirical index for the existence of absolute motion, and therefore for the existence of absolute space.

The first aspect can be understood in terms of Berkeley's critique of the existence of "abstract general ideas." General classificatory terms like "red," "man," and "triangle," are significant, but are not proper names or individual descriptions; that is there is no unique individual (a "universal" in re or in mente) that they name. They can be used, however, to refer to any individual in the class of red things, men or triangles. On the other hand the term "force" used to label an "occult" quality, or vital principle immament in matter, does not even have the distributive reference of legitimate classificatory terms; there is no individual entity referred to by the term "force." It is "inconceivable" in the sense of being unintelligible, from which it follows, of course, that "forces" are unobservable. The term "matter" or "material substance" is also unintelligible, either because the concept of "substratum" or "support" is unintelligible, or because the concept of an "*unthinking*" support of "ideas" or objects of sense is self-contradictory. Again the expression, "material substance," has, for Berkeley, no distributive reference.

[82] *Ibid.,* p. 233.
[83] Our comments are to some extend based on the point of view expressed by W. A. Suchting: "Berkeley's Criticism of Newton on Space and Motion," *op. cit.*

It is an interesting question whether Berkeley believes that those who posit the existence of universals, material substance and hylarchic principles in matter, have committed errors of reification based on some common activity termed "abstracting." In the first two cases, there appears to be some similarity. If we think of "abstracting" as removing or eliminating in thought what is unessential, and focusing on what is essential (focusing on the generic as opposed to the specific characteristics) than from a *functional* point of view we can "abstract" what is unessential in individual triangles, and focus on (predicate things of) the essential or generic properties. We leave open the question whether the etiology of our gaining concepts follows some such process. If we understand by "matter" not the "support of primary qualities" but rather a class term for "material bodies" (macroscopic or not), we can also ask what such bodies have in common, or what is essential to them that makes them "material" entities; whether it is extension alone, or extension and impenetrability, or extension, impenatrability and a vis inertia. The expression "matter" used this way, would, of course, have distributive reference; individual chairs, tables, electrons, photons, etc., would be material bodies. The problem of reification here, however, is quite different in a sense from the problem with respect to classificatory terms like "man," or "triangle." In fact we have a double problem here, based on Berkeley's conception of two different types of illegitimate abstraction: (a) there can be no universals *in re* or *in mente* referred to by the term "matter," simply because there can be no universals; (b) there can be no extended uncolered entities, because "color" and extension (at least visual extension) are inseparable; to separate them and claim the existence of extended uncolored entities would be an illegitimate reification.[84] The reification in (b) although distinct from (a) was quite important for Berkeley, for it represented to him what we can call the illegitimate mathematization of nature, the conversion of the metric properties of objects (the primary qualities) into the "real" properties of objects. Although Berkeley would agree, for example, that "color" does not enter into the geometric description of an extensive manifold, all extensive manifolds (again we would have to specify "visual" extension) must be colored. There is no entity referred to by the expression "extension simpliciter" nor "extension in itself."

The concept of "absolute space" can also be considered generated by

---

[84] See the excellent discussion of the different types of "abstraction" discussed by Berkeley in, Monroe Beardsly, "Berkeley on Abstract Ideas," printed in, *Locke and Berkeley*, ed. Martin and Armstrong, *op. cit.*, pp. 409-425.

a process of "abstraction," but again, two senses of this term must be distinguished. The first sense is expressed by the thought experiment of removing all material bodies from the universe, and asking what is left. If we say "space," we are viewing space as an entity distinct from and containing material bodies. And the relation of container to contained is not that of genus to species.

The second sense is expressed in the question: what remains, if I remove from my conception of "body" properties such as color, taste, solidity (or impenetrability), and following Newton remove "mass" as well? The Cartesian answer, accepted and expressed by Newton in an unpublished work, was that "extension" or "spatiality" remains, "the uniform and unlimited stretching out of space in length, breadth and depth." [85] Here spatiality or extensiveness can be said to stand as genus to the qualitatively different types of space. One might argue, for example, that a yellow and a green patch qua extended, have certain properties in common, their geometrical properties (understood in the general sense of both patches having an Euclidean structure).

Berkeley often (particularly in the sections of *De Motu* and the *Principles* we are dealing with) focuses his criticum on what he considers to be the reification having to do with the first "sense" of the concept of space. For example, *Principles* 116:

And perhaps if we inquire narrowly, we shall find we cannot even frame an idea of *pure space* exclusive of all body. This I must confess seems impossible, as being a most abstract idea.

Or *De Motu* 53 and 54:

And so let us suppose that all bodies were destroyed and brought to nothing. What is left they call absolute space, all relation arising from the situation and distances of bodies being removed together with the bodies. Again, that space is infinite, immovable, indivisible, insensible, without relation and without distinction. That is, all is attributes are privative or negative. (*De Motu* 53)
Such an idea, (absolute space) moreover when I watch it somewhat more intently, I find to be the purest idea of nothing, if indeed it can be called an idea. (*De Motu* 54)

The concept of "space" in this sense is not self-contradictory, but vacuous, in the sense that there is nothing in our sensible experience to which the

[85] Isaac Newton, *Unpublished Papers of Isaac Newton*, ed. E. A. Robert Hall and Marie Boas Hall, (Cambridge: Cambridge University Press; 1962), "De Gravitatione and Aequipondio Fluidorium." ("On the Gravity and Equilibrium of Fluides.") p. 133.

term applies. Berkeley's observational demand, however, is too stringent; we may not be able to "imagine space devoid of body," yet extensiveness or spatiality is a characteristic of the visual field, and therefore an object of perceptual experience.

Berkeley entertains the possibility (*Principles* 116, *De Motu* 55) that we have an "idea" of space, through the kinaesthetic sensation of moving our limbs without hindrance:

> When, therefore, supposing all the world to be annihilated besides my own body, I say there still remains *pure space*, thereby nothing else is meant but only that I conceive it possible for the limbs of my body to be moved on all sides without the least resistance; but if that, too were annihilated, then there could be no motion and consequently no space. Some, perhaps may think the sense of seeing does furnish them with the idea of pure space; but it is plain from what we have elsewhere shown, that the ideas of space and distance are not obtained by that sense. See the *Essay Concerning Vision*. (*Principles* 116)

The passage appears to confuse, the issue of whether we have any "idea" of space, with the issue of whether we have any measure or index of absolute space. The issue of the "idea" of space is really irrelevant to question of whether there exists a physical envelope separable from and containing physical bodies and processes. That I have an "idea" of a table is not an argument for Berkeley that independent of minds, tables exist. If part of the meaning of the expression "absolute space" (like "material substance") is that what is purportedly refers to exists independently of minds, then clearly absolute space cannot be an "idea." Interestingly, this point is recognized in the *Three Dialogues*. Berkeley (Philonous) says:

> But allowing that distance was truly and immediately perceived by the mind, yet it would not thence follow it existed out of the mind. For whatever is immediately perceived is an idea; and can any idea exist out of the mind? (*Dial.* 1 p. 143)

The passage demonstrates Berkeley's recognition that the argument of the *Essay*, that "distance" is not immediately perceived, is not an argument for immaterialism. The point is that "ideas" cannot exist unperceived; the same would hold true for any "idea" of space.

Of more interest in *Principles* 116 is how "motion" is relevant to the issue of the existence of absolute space. Berkeley's claim that without "motion" there could be no "space," is ambiguous. What he most likely means is that we "conceive" space by means of the motion of our limbs; and that (assuming all other bodies were previously removed) we would no longer have the "idea" of the separation of one body (our hand) from another (our chest) if we too were annihilated. The argument could be put this way: the thought experiment by which we imagine ourselves alone in the universe, moving our limbs, merely demonstrates that our cognizance of space is generated through an awareness of the motion of one body relative to another. If we continue the thought experiment and annihilate even this relative motion between one part our body and another, we can have no cognizance of motion and therefore of space.

We are sometimes deceived by the fact that when we imagine the removal of all other bodies, yet we suppose our own to remain ... we imagine the movement of our limbs fully free on every side ... but motion without space cannot be conceived. None the less if we consider the matter again we shall find, 1st relative space conceived, defined by the parts of our body; 2nd a fully free power of moving our limbs obstructed by no obstacle; and besides these two things, nothing. (*De Motu* 55)

We are interested in Berkeley's first point, that the thought experiment allows us to conceive only "relative space." The suggestion is that "relative space," exhausts the meaning of the term "space," and that by "relative" is meant the situation of any body with respect to (relative to) other bodies. All we can say concerning the movement of my arm (given the annihilation of the remainder of the material universe) is that it changes its position with respect to some other material object (some other part of my body). The broader issue, then, is whether there is some sensible criterion that would allow us to apply the expression "absolute motion," for if this were the case, the frame of reference for such a motion would be absolute space.

We should remark at this point that the term "conceive," as used here, seems clearly to mean "sense" or "imagine." Even Newton would most likely agree that "absolute space" cannot be "conceived," in this sense.[86] However, for Newton, the concept of absolute space is not only intelligible, but dynamically required as the frame of reference for "absolute" accelerative motions, for which, he believed, there was empirical evidence.

We will consider below Berkeley's evaluation of this "evidence." Before this, however, it is interesting to note that Berkeley entertains (and rejects) the possibility that there is one positive property that might be

---

[86] Newton, *Principia, op. cit.*, see the Scholium to Definition VIII, pp. 8-13; "And so, instead of absolute places and motions, we use relative ones; and that without any inconvenience in common affairs, but in philosophical disquisitions, we ought to abstract from our senses, and consider things themselves, distinct from what are only sensible measures of them. For it may be that there is no body really at rest, to which the places and motions of others may be referred." (p. 10)

attributed to "absolute space," and that is "extension." His criticism is that we have no "idea" of extension *simpliciter*, where "idea" means "object of sense."

But what sort of extension, I ask, is that which cannot be divided, nor measured, no part of which can be perceived by sense or pictured by the imagination? For nothing enters the imagination which from the nature of the thing cannot be perceived by sense, since indeed the imagination is nothing else than the faculty which represents sensible things either actually existing or at least possible. Pure intellect, too, knows nothing of absolute space. That faculty is concerned only with spiritual and in extended things, such as our minds, their states, passions, virtues and such like. (*De Motu* 53)

The issue involved has some historical interest. What Berkeley is objecting to is the tradition – we can call it Cartesianism – which identifies the generic properties of perceived extensive magnitudes with the "real" or objective properties of that alleged container of physical processes, called "absolute space." And the generic or structural properties of this space (properties which all "figures" qua extensive magnitudes, share), are said to be articulated by Euclidean geometry. The two senses of "abstraction" we previously mentioned appear, in this tradition, to have coalesced. That which remains after I "remove" all bodies from the universe is identified with what remains after I remove from my conception of "body" all non-extensional predicates like color, solidity, inertia, etc.

From this point of view, Euclidean geometry becomes the science, not of the physical boundaries of objects conceptually but not physically separable from the objects themselves, but rather the science of pure space itself. There were perhaps two essential reasons for this shift in the conception of geometry. The first was that classical physics, culminating in the Newtonian system, emphasized the application of mathematics to the study of motion. Moving bodies, secondly, traverse spatial "paths," and these paths appear to be required by the "laws" of mechanics to have certain properties. For example a single body (according to Newton) moving only according to the first law of motion, would describe a rectilinear path, or what we can call a Euclidean geodesic. Given the success of the Newtonian system, one might claim that we have a good deal of evidence that space is "Euclidean."

However, why not merely say that the "space" of our experience is Euclidean, rather than that "absolute space" is Euclidean? One could argue that even if one holds a relational theory of "space" that "space" means no more than, to quote Leibniz, a "certain order of coexistence" of bodies; [87] we can apprehend the structure of this order to be Euclidean. In fact, two senses of the term "absolute" have to be distinguished: what we can call a "metaphysical" sense, and a "dynamical" sense. And these senses correspond to the two types of "abstraction" previously mentioned.

"The "metaphysical" sense of "absolute space" connotes the view that the generic properties of "spaces" (the structure of extension itself) constitute the only "real" properties of extensive magnitudes; whereas the non-extensive properties like color or solidity (viewed as a tactile sensation) are modifications of the perceiving subject, not properties of extended objects themselves. Side-stepping the issue for the present, of how one is said to know that the structure of space is Euclidean, we can agree with Berkeley that within the context of his usage of the term "idea," we have no "idea" of extension simpliciter, just as we have no "idea" of "material substance" as substratum devoid of its properties.

On the other hand, even if we allowed that the generic properties of "spaces" (by "spaces" we mean again yellow or red patches, etc.) satisfy the Euclidean axioms, and allowed the hypostatization that these "generic properties" were the only "real" properties of extended objects (perhaps adding others like "mass"), and moreover allowed that in some sense of the term "idea" (perhaps in a Platonic sense) this geometrical structure was an "idea," we still would not have established "absolute space" in the "dynamic" sense as the frame of reference for either purely inertial motion (that is a body acting in accordance with its inherent power – (its vis insita) – to resist changes in motion), or for certain "rotational" motions said to produce centrifugal forces.

Our own view is that the problem of the "idea" of extension is best separated from the epistemological and metaphysical framework in which it is placed by Berkeley, and discussed in terms of the question of what is the "object" of geometry. We will take up the issue when we deal with Berkeley's philosophy of mathematics.

We will now turn to Berkeley's criticism of Newton's belief that certain dynamical evidence compels us to assert the existence of absolute space. Newton, as is well known, argued that although we could not establish "absolute space" as a frame of reference for a body moving at uniform velocity, we could establish "absolute space" as the frame of reference for certain rotational motions, those attendant with centrifugal forces.[88]

---

[87] Quoted by Jammer, *Concepts of Space, op. cit.,* p. 115. Jammer quotes from Leibniz's fifth letter to Clarke.

[88] Newton, *Principia, op. cit.,* Law (Axiom) III, Corollary V. "The motions of bodies included in a given space are the same among themselves, whether that motion

In essence Newton's argument is that certain deformations in material objects indicate the presence of centrifugal forces, and hence the absolute acceleration of the body deformed. By merely "internal" observation (that is, without reference to any other co-existing bodies) of such deformations, we can conclude that the body is in motion absolutely and not merely relatively in terms of a translation with respect to some other body judged to be fixed.

Newton offers two examples in the "Scholium" to definition VIII, Book I of the *Principia*, as evidence for his view. The first has to do with the tension developed in a chord, produced by whirling two globes around a common center; the second is the more famous "bucket experiment," where he demonstrates that when a bucket of water is rotated, there will be, after a while, a deformation in the water's surface (it will assume a concave shape), a deformation present when the water is both at rest and in motion *relative* to the bucket.[89] Newton's conclusion is as follows:

The ascent of the water shows its endeavor to recede from the axis of its motion; and the true and absolute circular motion of the water, which is here directly contrary to the relative, becomes known and may be measured by this endeavor.[90]

It is interesting to note that in an earlier unpublished work, Newton entertains the hypothesis, only to reject it, that relative motion, for example, relative to the fixed stars, is sufficient to account for centrifugal forces. Speaking of such forces that allegedly accompany the earth's rotation, he comments:

As if it would be the same, whether with a tremendous force, he (God) should cause the skies to turn from East to West, or with a small force turn the earth in the opposite direction. But who will imagine that the parts of the earth endeavor to recede from its center on account of a force impressed only on the heavens? Or is it not more agreeable to reason that when a force imparted to the heavens makes them endeavor to recede from the center of the revolution thus caused, for that reason it is the sole body properly and absolutely moved, although there is the same relative motion of the bodies

is at rest, or moves uniformly forwards in a right line without any circular motion."
p. 19

If a system (S) is an inertial system (one in which the Newtonian laws are true) there is no way of detecting whether (S) is at rest or in uniform motion with respect to absolute space. In all such systems, (assuming the invariancy of mass with respect to velocity) if a force produces an acceleration (g) in one system, will produce an acceleration (g) in any system moving uniformly with respect to the first. Therefore changes in momentum, which is the measure of forces will be the same in all such systems.

[89] Newton, *Principia, op. cit.*, Scholium to Def. VIII, p. 11-12.
[90] *Ibid.*

in both cases. And this physical and absolute motion is to be defined from other considerations then translation, such translation being designated merely as external.[91]

The passage expresses a number of points, all of which, we believe, are explicitly or implicitly maintained in the *Principia*: (a) impressed forces are distinct from, and causes of "motion"; (b) material deformations are the effect of impressed forces; and (c) the impressed force, and hence the change in momentum, must be predicated of the body in which the material deformation takes place. The *Principia* clearly asserts (b): [92] that the distinction between "true" and "relative" motion is in terms of "impressed forces." Therefore, (c) appears eminently reasonable to Newton; why should the deformations take place in the body that is not moved? The importance of (c) is that it reminds us that when Newton entertains the possibility that "relative motion" is sufficient to explain centrifugal forces, he is not entertaining the possibility that the concept of "true" motion is meaningless. He is merely considering the possibility that the body undergoing the "true" motion is not the body in which the effects of centrifugal forces occur.

Berkeley's critique of Newton's arguments for absolute motion, and hence absolute space as its frame of reference, is found both in the *Principles* and in *De Motu,* though there are important differences in the two accounts.

In the *Principles,* after quite accurately paraphrasing Newton's distinction between absolute and relative space, time and motion,[93] (*Principles* 110-111) Berkeley comments that:

not withstanding what has been said, it does not appear to me that there can be any motion other than *relative*; so that to conceive motion there must be at least conceived two bodies, whereof the distance or position in regard to each other is varied. Hence if there was only one body in being it could not possibly be moved. This seems evident, in that the idea I have of motion does necessarily include relation. (*Principles* 112)

Suchting suggests that by "conceive" Berkeley means "imagine," and therefore without further assumptions (for example, immaterialism), Berkeley's conclusion, that "if there was only one body in being, it could not possibly be moved," does not logically follow.[94] In our view, the "semantics" of "conceive" for Berkeley is more complicated. Taking the

[91] Hall, *Newton's Unpublished Works, op. cit.,* p. 128.
[92] Newton, *Principia, op. cit.,* p. 11.
[93] *Ibid.*
[94] Suchting, *op. cit.,* p. 189.

passages in the *Principles,* together with those in *De Motu,* one gains the impression that "conceive" may have the stronger sense of "logically possible," as opposed to "imaginable," and consequently that "inconceivable" may have the sense of "logically impossible." This suggests that Berkeley may be explicating the *meaning* of the term "motion" as opposed to merely discussing the perceptual requirements for the apprehension of motion. Support for this view comes from some of the relevant passages in *De Motu:*

... We must point out that no motion can be understood without some determination or direction, which in turn cannot be understood unless besides the body in motion, our own body also, or some other body, be understood to exist at the same time. For up, down, left, right and all places and regions are founded in some relation, and necessarily connote and suppose a body different from the body moved. (*De Motu* 58)

Here the clear suggestion is that it is part of the meaning of the expression, "X is in motion," that there exists some other body in respect to which the position of X is changing. In the next section, both senses of the term "conceive" appear to be involved.

Then let two globes be conceived to exist and nothing corporeal besides them. Let forces then be conceived to be applied in some way; whatever we may understand by the application of forces, a circular motion of the two globes round a common center cannot be conceived by the imagination. Then let us suppose that the sky of the fixed stars is created; suddenly from the conception of the approach of the globe to different parts of that sky the motion will be conceived. This is to say that since motion is relative in its own nature, it could not be conceived before the correlated bodies without correlates. (*De Motu* 59)

On the one hand, "conceive," as used in the passage, clearly means "imagine." We cannot imagine the globes (assuming the distance between them kept constant) moving except with respect to some other body. On the other hand, Berkeley asserts that "motion is relative in its own nature," suggesting it is part of the meaning of "motion" that a relation between bodies is involved. There is no incompatibility here; it is because, as Berkeley suggests, the meaning of "motion" involves relation that "absolute motion" cannot be "conceived" ("imagined"). The point is repeated later in *De Motu:*

No motion can be recognized or measured unless through sensible things. Since then absolute space in no way affects the senses, it must be quite useless for the distinguishing of motions. Besides, determination or direction is essential to motion; but that consists in relation. Therefore it is impossible that absolute motion should be conceived. (*De Motu* 63)

Newton would have agreed with the first statement of the passage; it is the observable tension in the chord connecting the two globes, or the observed concavity of the water's surface, that gives evidence for the existence of absolute rotations. The second part of the passage suggests, however, that the issue is more than the recognition or measurement of motion, and concerns the meaning of the concept of motion itself. If, in fact, as the passage suggests, it is part of the meaning of the expression, "(X) is in motion," that (X) has changed its situation with respect to some other body, then motion with respect to absolute space is a logical impossibility. From which it follows that such motion is not "conceivable," in the sense of not being "imaginable."

However, the question remains, how does Berkely handle Newton's claim that we have evidence of a purely internal nature for absolute motion, evidence which requires no comparison of the "moving" body with any other body in the universe? Newton, if we follow the remarks in his unpublished essay, found it implausible that terrestrial manifestations of centrifugal forces could have been produced by a rotation of the fixed stars. The earth, then, must truly be rotating. And since all motion is a change in the situation of the "moving" body with respect to something else, the earth must rotate with respect to absolute space.

What is quite interesting is that Berkeley, like Newton, does draw a distinction between "true" and "*apparent*" motion (in this context, the term "apparent" rather than "relative," best expresses Berkeley's meaning). It is not enough to denominate a body "moved," that it has undergone a change of place with respect to some other body. An equally necessary condition is that some "impressed force" has been applied to the body said to be in motion.

But, though in every motion it be necessary to conceive more bodies than one, yet it may be that one only is moved, namely, that on which the force causing the change of distance is impressed, or, in other words, that to which the action is applied. For, however some may define relative motion, so as to term that body 'moved" which changes its distance from some other body, whether the force or action causing that change were applied to it or no, yet as relative motion is that which is perceived by sense, and regarded in the ordinary affairs of life, it should seem that every man of common sense knows what it is as well as the best philosopher. (*Principles* 113)

Berkeley, unfortunately, does not make clear how the sense of "force" expressed in the above passage relates to other discussions of the term in *De Motu* and the *Principles*. On the surface, at least, it appears to be quite incompatible with them. "Force" here has the sense of efficient cause, and is not meant to be elliptical for "mathematical hypothesis." What is interesting also, is that Berkeley, much like Newton, appeals to an intuitive sense of plausibility to make his point:

Now I ask anyone, whether, in his sense of motion as he walks along the streets, the stones he passes over may be said to move, because they change distance with his feet? (*Principles* 113)

In addition, it is important to note that Berkeley, like Newton, when he entertains the notion that "relative" motion is sufficient to explain certain effects, is not suggesting that the concept of "real" as opposed to "apparent" motion is meaningless. Rather, entertaining such a notion means no more than considering the possibility that either the body (A) or (B) in relative motion can be considered "really" moving.

In a related section Berkeley clarifies what he considers to be the conditions sufficient to say that a body is in motion:

first that it change its distance or situation with regard to some other body, and secondly that the force or action occasioning that change be applied to it. (*Principles* 115)

With respect to the second condition, it appears identical to Newton's assertion in the scholium to definition VIII Book I of the *Principia,* that:

The causes by which true and relative motions are distinguished, one from the other, are the forces impressed upon bodies to generate motion.[95]

What, then, constitutes Berkeley's disagreement with Newton? If we turn to his comments on Newton's "bucket experiment," we are not aided very much in gaining an answer. In the *Principles,* Berkeley contends:

As to what is said of the centrifugal force, that it does not at all belong to circular relative motion; I do not see how this follows from the experiment which is brought to prove it. (See *Philosophiae Naturalis Principia Mathematica, in Schol. Def. VIII*) For the water in the vessel at that time wherein it is said to have the greatest relative circular motion, has, I think, no motion at all; as is plan from the foregoing section. (*Principles* 114)

We agree with Suchting, that by "the foregoing section," Berkeley means section 113, where it is claimed that a necessary condition to denominate a body "moved" is that it be the subject of an impressed force.[96] The passage, however, misinterprets Newton. By "relative" motion Newton means no more than change of place with respect to some other body. Newton does not divide "relative" motions into "true" and "apparent"

[95] Newton, *Principia, op. cit.,* p. 11.
[96] Suchting, *op. cit.,* p. 193.

as does Berkeley (we will discuss this below). In addition, the passage fails to come to grips with Newton's major contention, that the deformation in the surface of the water is evidence of the existence of impressed forces on the water, and hence, evidence for the "absolute" acceleration of the water. In *De Motu*, the two references to the "bucket" experiment respectively misuse the notion of the composition of forces, and raise what appears to be an irrelevant issue concerning the "circularity" of the bucket's motion:

As regards circular motion many think that, as motion truly circular increases, the body necessarily tends ever more and more away from its axis. This belief arises from the fact that circular motion can be seen taking its origin, as it were, at every moment from two directions, one along the radius and the other along the tangent, and if in this latter direction only the impetus be increased, then the body in motion will retire from the centre, and its orbit will cease to be circular. But if the forces be increased equally in both directions the motion will remain circular, though accelerated – which will not argue an increase in the forces of retirement from the axis, any more than in the forces of approach to it. Therefore we must say that the water forced round in the bucket rises to the sides of the vessel, because when new forces are applied in the direction of the tangent to any particle of water, in the same instant new equal centripetal forces are not applied. From which experiment it in no way follows that absolute circular motion is necessarily recognized by the forces of retirement from the axis of motion. (*De Motu* 60)

Berkeley too considers the experiment (as Suchting correctly recognizes in a footnote),[97] only in terms of that aspect when the water is *acquiring* a uniform rim velocity and a certain deformation of its surface. During this time it may be meaningful to speak of a "new" impetus along the tangent, since there is an increase in the velocity component along the tangent (thus an acceleration) until uniform rim velocity has been achieved. However, as Suchting indicates, the problem is that the deformation remains after uniform rim velocity has been achieved, when, within the context of the Newtonian laws, the only "force" (measured by a change in momentum) is centripetal. From Newton's point of view, the *continued* deformation of the water is a measure of the "resistance" of the water particles to any change in their "normal" (rectilinear) path, and hency of their circular motion.

Although Suchting is correct in his criticism of Berkeley's conception of tangential forces, the criticism is somewhat misdirected.[98] The latter's claim in this section, as well as the following two, is that we cannot

---

[97] *Ibid.,* p. 195.
[98] *Ibid.*

"necessarily" argue from the deformation of the water to its "absolute circular motion." Now Berkeley's contention could mean one of three things: (A) there must be some non-uniform motion of the "body" deformed, but it need not be circular; (B) such deformations do not necessarily entail any motion of the body deformed; and (C) motions are entailed by such deformations, but it is meaningless to refer to absolute space as a frame of reference.

Section 62 (*De Motu*) appears to argue for (A). Leaving open the question whether there is or is not absolute space, Berkeley comments:

We must not omit to point out that the motion of a stone in a sling or of water in a whirled bucket cannot be truly be called circular motion as that term is conceived by those who define the true places of bodies by the parts of absolute space, since it is strangely compounded of the motions, not alone of bucket or sling, but also of the daily motion of the earth around its own axis, of her monthly motion about the common centre of gravity of earth and moon, and of her annual rotation around the sun. And so that account each particle of the stone or the water describes a line far removed from the circular.

Suchting argues that although correct, Berkeley's point is irrelevant to the purpose of "Newton's experiment":

which was to provide an example of an action of forces, and hence an acceleration, which could not be explained by reference to relative motions between material bodies. It does not matter for this purpose whether the orbits of the water particles are exactly circular, for so long as they pursue a non-uniform or non-rectilinear path there is force and hence acceleration.[99]

Suchting's last statement is misleading; it is not the non-uniformity of the path of a water particle which is sufficient evidence of the action of forces; rather it is that the deformation of the "water" (not the water particle) which is evidence for the existence of forces and therefore for the water's acceleration. In addition, to speak of the "correctness" of Berkeley's point about the non-circularity of the water's path presupposes some conception of a frame of reference for the composite motion he speaks of. Newton himself never claimed that "absolute space" could be used (we cannot select a point at rest in absolute space) for the determination of motions. Therefore the claim that, for example, the water has a non-circular motion with respect to the sun, would not necessarily argue against its circular motion with respect to absolute space.

However, (B) is perhaps a more radical proposal: that material deformations (like the surface of the water in Newton's bucket) entail the

[99] *Ibid.*

existence of impressed forces, but do not entail that the deformed body is in motion.[100] This is compatible with the view expressed in the *Principles,* that being under an impressed force is a necessary but not a sufficient condition for the attribution of motion to a body. Although *De Motu,* given its more searching analysis of the concept of "force," does not stress this point, it is somewhat hesitantly repeated:

For however forces may be impressed, whatever conations there are, let us grant that motion is distinguished by action exerted on bodies; never, however, will it follow that space, absolute place, exists, and that change in it is true place. (*De Motu* 64)

Such a view might explain the interesting passage from the *Philosophical Commentaries:*

I differ from Newton in that I think the recession ab axe motus is not the effect or index or measure of motion, but of the vis impressa. It showeth not what is truly moved but what has the force impressed on it. Or rather that which hath an impressed force. (*Commentaries* 456)

However, to a defender of Berkeley, proposal (B) might appear to beg the question concerning the existence of "absolute space." If the body suffering the deformation is said to be at "rest," how is this term understood? If it means with respect to absolute space, the question is begged; if it means with respect to some physical frame of reference, it seems obvious that with respect to some frames, the body is at rest, and with respect to others it is in motion. Putting aside this difficulty for the moment (the choice of a reference frame) we might take Berkeley to be expressing proposal (C): that deformations such as the concavity of the water's surface in the bucket do entail motion with respect to some reference frame other then absolute space. From this perspective, we could emend the passage in the *Commentaries* previously cited, to read:

I differ from Newton in that I think the recession ab axe motus is not the effect or the index of motion [with respect to absolute space]...[101]

This interpretation of the passage is compatible with the view that such "recession" is an "index" of motion with respect to some material reference frame, for example, the fixed stars.

---

[100] Newton's "Second Law" strictly entails that an accelerated motion requires the presence of an "impressed force." It would not logically follow that the presence of an "impressed force" entailed the acceleration of the body upon which the force was impressed. In a situation of dynamic equilibrium (a book resting on a table) an object is subject to impressed forces without accelerating. Berkeley appears to suggest in *De Motu* that it is meaningless to speak of the presence of "forces" in such situations.

[101] *Philosophic Commentaries* (456).

It is important to put in perspective what, in fact, is Berkeley's disagreement with Newton. For Berkeley, "relative motion" and "real motion" are compatible expressions. All motion, in his view, is relative, in the sense that it is part of the meaning of the expression "(X) is moved," that (X) changes its situation (distance) with respect to some other body. Within the class of relative motions, however, we have "real" motions, where the body said to be in motion has been acted on by an impressed force, and "apparent" motions, where this is not the case. Newton distinguishes "relative" from "absolute" motion; the latter connotes not only that the body said to be in motion acts under an impressed force, but that its motion must be referred to absolute space as a frame of reference. To allow that motion was merely "relative" implied not that the concept of absolute motion was devoid of meaning, but that either of the bodies in relative motion could legitimately be said to be the one "truly" moved. Both men would probably have agreed that the "bucket" rather than the "fixed stars" was truly in motion, and both would have agreed that the deformation of the water's surface indicated the existence of an impressed force on the water.

Berkeley's attempt, however, to combine a belief in the essential relativity of motion with a distinction between "real" and "apparent" motion would appear to have considerable difficulties *if* he wishes to maintain that certain material deformations are not only an index of impressed forces but of some sort of motion in the body on which the force is impressed.

The difficulty can be expressed this way: *given* a situation (which was also the situation for Newton) where the bucket has a relative rotation with respect to the earth or the fixed stars, Berkeley would choose to say the bucket rather than the fixed stars is really rotating, since it (the contained water) suffers a deformation. But why choose the fixed stars as a privileged reference frame. With respect to other possible frames (one, for example, having the same angular velocity with respect to the fixed stars, as the bucket) the bucket would appear to be at rest. However, with respect to such frames we would have no relative, and therefore from Berkeley's point of view, no real motion, although there would be the same deformation of the water.

Newton, in the essay earlier referred to, wrestled with this problem, and his remarks are worth quoting:

For unless it is conceded that there can be a single physical motion of any body, and that the rest of its changes of relation and position with respect to other bodies are so many external designations, it follows that the Earth (for example) endeavors to recede from the centre of the sun on account of a motion relative to the fixed stars, and endeavors the less to recede on account of a lesser motion relative to Saturn, and the aetherial orb in which it is carried, and still less relative to Jupiter and the swirling aether which occasioned its orbit, and also less relative to Mars and its aetherial orbit . . . and indeed relative to its own orb it has no endeavor, because it does not move in it. Since all these endeavors and non-endeavors cannot absolutely agree, it is rather to be said that the only motion which causes the Earth to endeavor to recede from the Sun is to be declared the Earth's natural and absolute motion.[102]

Newton's point here seems to be as follows: if we assume that the magnitude of centrifugal forces, measured by certain material deformations, is proportional to a change in the momentum of the body in which those deformations occur, then we are not at liberty to choose any reference frame we like in analyzing the motion of this body. There is, for Newton, a privileged reference frame, with respect to which his laws of motion are true, and this frame is to be identified with absolute space.

Berkeley, we believe, would agree with the above position, with the significant substitution of the fixed stars for absolute space as the privileged frame of reference. It would, however, be going beyond the sense of the admittedly condensed Berkeleian texts to say that his view on this issue is comparable to the "conventionalism" of Ernst Mach. It is more plausible to say that for Berkeley, the fixed stars constitute a privileged frame of reference, since using them reveals what in fact *is* the case, that (as expressed by Newton's second law) the magnitude of impressed forces is proportional to changes in momentum of the body on which those forces are impressed.

Take, for example, the rather "conventionalist" sounding passage in *De Motu*:

Further, since the motion of the same body may vary with the diversity of relative place, nay actually since a thing can be said in one respect to be in motion and in another respect to be at rest, to determine true motion and true rest, for the removal of ambiguity, and for the furtherance of the wider view of these philosophers who take the wider view of the system of things, it would be enough to bring in, instead of absolute space, relative space as confined to the heavens of the fixed stars considered at rest. But motion and rest marked out by such relative space can conveniently be substituted in place of the absolutes which cannot be distinguished from them by any mark. (*De Motu* 64)

---

[102] Hall, *op. cit.*, p. 127. On the relation of Newton's earlier to his later published work, see: E. W. Strong, "Barrow and Newton," *Journal of the History of Philosophy*, Vol. VII No. 2, April 1970, pp. 155-172.

Although the passage suggests that the choice of reference frame may be, in part, due to convenience, the phrase "to determine true motion and true rest," cannot be considered anomalous. Rather, the phrase is consistent with Berkeley's view that "true" motion can be distinguished from "apparent" motion in terms of which body is "acted" upon by impressed forces. In this context "impressed forces" cannot be identified in meaning with change in momentum with respect to the fixed stars. (That is F cannot *mean* MA, where A (acceleration) is measured using the fixed stars as a frame of reference.) If this were the case, Berkeley's claim would be equivalent to the tautology that using the fixed stars as a frame of reference reveals those bodies whose momentum changes with respect to the fixed stars. If the fixed stars used as a reference frame are to be said significantly to reveal "true motion and rest," where the latter is predicated of bodies which are (or are not) acted upon by impressed forces, then impressed forces must be defined in other terms then mere change in momentum with respect to the privileged frame of reference.

The above discussion reveals what is ultimately problematical with *De Motu,* taken as a whole, its ambiguity concerning the status of "impressed forces" in the Newtonian system. Alternatively stated, there is a central ambiguity concerning how Berkeley understands Newton's first two laws (axioms) of motion. At times he suggests they are empirical generalizations of a rather straightforward sort. Yet if this were true, "impressed force" would have to be conceptually distinguished from "change in momentum." At other times he suggests that the legitimate use of "forces" in mechanics is comparable to using a mathematical fiction, like the "lines and angles" of geometers. This view is often apparent in his discussion of the "composition of forces." For example, section 61 in *De Motu*: [103]

A curve can be considered as consisting of an infinite number of straight lines, though in fact it does not consist of them. That hypothesis is useful in geometry; and just so circular motion can be regarded as arising from an infinite number of rectilinear directions – which supposition is useful in mechanics. Not on that account must it be affirmed that it is impossible that the centre of gravity of each body should exist successively in single points of the circular periphery, no account being taken of any rectilineal direction in the tangent or the radius.

The passage appears to suggest that from a physical point of view, circular motion might be "natural motion"; that is, require no explanation in terms of impressed forces; but that to consider the motion of the moving

[103] See the comparable discussion of the "parallelogram of forces" in *De Motu* (18).

body at every moment as the resultant of two vector quanitities (an acceleration towards the center and a uniform velocity along the tangent) is merely a useful device. Yet as Huygens had shown and Newton demonstrated in the *Principia,* a body in circular motion has an acceleration towards the center.[104] In any case circular motion is non-rectilinear and according to the first law must be motion under an impressed force. In sum, then, the achievement of the program of eliminating in mechanics forces as efficient causes, requires the kind of radical re-thinking of Newton's laws of motion which we find in Mach, but not in Berkeley's *De Motu.*

We would note that if anything in the Newtonian system could be conceived as an "ideal" entity or mathematical fiction, it should have been, from Berkeley's own point of view, "absolute space" and not the Newtonian "impressed force." One might think of absolute space as that ideal reference frame for which Newton's laws hold absolutely, as opposed to all material references frames (including the fixed stars) for which the laws may only hold approximately.[105] The concept of "absolute space" might then have some comparability to mass points exercising attractive powers, which Newton does entertain as a mathematical hypothesis.

We will conclude this discussion of *De Motu,* by commenting in more detail on the thesis that Berkeley's critique of Newton's concepts of absolute space, time and motion anticipates the work of Ernst Mach. This view is confidently expressed, for example, by Popper. "What is perhaps most striking," he says,

is that Berkeley and Mach both great admirers of Newton, criticize the ideas of absolute time, absolute space, and absolute motion, on very similar lines. Mach's criticism exactly like Berkeley's culminates in the suggestion that all arguments for Newton's absolute space (like Foucault's pendulum, the rotating bucket of water, the effect of centrifugal forces upon the shape of the earth, fail because these movements are relative to the fixed stars.[106]

A comparable view is put forth by R. H. Dicke in his essay "The Many Faces of Mach." Dicke refers to Berkeley's contention in *De Motu* (Sec. 59) that the motion of two globes whirled around a common center cannot be conceived until we imagine the fixed stars as a frame of reference; and claims that this is an early statement of what has been later formulated as "Mach's Principle." Dicke quotes from Mach's Science of Dynamics:

[104] Newton's demonstration in *Principia,* Book I, Section II, Proposition 2. Theorem 2, *op. cit.,* p. 34.
[105] Nagel, *op. cit.,* p. 212.
[106] Popper, *op. cit.,* p. 445.

For me only relative motions exist . . . when a body rotates relative to the fixed stars, centrifugal forces are produced; when it rotates relatively some different body not relative to the fixed stars, no centrifugal forces are produced. I have no objection to calling the first, rotation, as long as it be remembered that nothing is meant except relative rotation with respect to the fixed stars.[107]

A closer look at Mach's position, however, discloses that he differs significantly with Berkeley. Mach's view essentially is that the choice of a reference frame is dictated solely by reasons of convenience. For example, we note the following passage from the *Science of Mechanics:*

But if we take our stand on the basis of facts we shall find we have knowledge only of relative spaces and motion. Relatively, not considering the unknown and neglected medium of space, the motions of the universe are the same whether we adopt the Ptolmaic or the Copernican point of view. Both views are, indeed equally correct; only the latter is more simple and more practical. The universe is not twice given, with an earth at rest and an earth in motion; but only once, with its relative motions alone determinable. It is, accordingly, not permitted us to say how things would be if the earth did not rotate. We may interpret the one case that is given us in different ways.[108]

The very thought experiment which conceived fixing the bucket and rotating the fixed stars begs the question since it assumes the significance of the distinction in question, between absolute and relative motion.

. . . the system of the world is only given once to us, and the Ptolmaic or Copernican view is our interpretation, but both are equally actual. Try to fix Newton's bucket and rotate the heaven of fixed stars and them prove the absence of centrifugal forces.[109]

In this view, centrifugal forces are manifestations of *relative rotations,* and no distinction between real and apparent motion dictates the choice of a reference frame.

Newton's experiment with the rotating vessel of water simply informs us that the relative rotation of the water with respect to the sides of the vessel produces no noticeable centrifugal forces, but that such forces are produced by its relative rotation with respect to the mass of the earth and the other celestial bodies.[110]

On the other hand, Berkeley evidently distinguishes between "real" and "apparent" motion (merely relative), the former being predicated of

[107] R. H. Dicke, "The Many Faces of Mach" (H. Y. Chiu and W. F. Hoffman (eds.) *Gravitation and Relativity* (New York: Benjamin, 1964), p. 122.
[108] Mach, *op. cit.,* pp. 283-284.
[109] *Ibid.,* p. 279.
[110] *Ibid.,* p. 284.

bodies under the influence of impressed forces. There is little, if any evidence in *De Motu* that Berkeley would accept *mere* relative rotation of the bucket and the fixed stars as causally antecedent to the existence of centrifugal forces in the former. Mach, on the other hand, would claim that regardless of the decision about a reference frame, it is only the relative rotation of bucket and stars that can be meaningfully spoken of as the causal antecedent to the deformation of water in the bucket.

The difference could be alternatively expressed by saying that whereas Berkeley, like Newton, finds it meaningful but implausible to attribute the deformation of the water to a *real* rotation of the fixed stars, Mach would find the very conception of such a possibility to be meaningless. For him, the denomination "real," with respect to the motion of the water, has no more conceptual content than "motion with respect to the privileged reference frame"; whereas for Berkeley it has the antecedent meaning of "motion that is a consequence of an impressed force."

Concerning the issue of "absolute motion" as well as the issue of "force," the lack of clarity in *De Motu* on the status of impressed forces makes it difficult to consider Berkeley in any substantial sense a precursor of Mach. The relativization of motion achieved by Mach required not only dispensing verbally with impressed forces as efficient causes, but restructuring the first and second Newtonian laws of motion in light of this program to eliminate "forces" from mechanics. For example, we know that for Mach, the second law has the status of a definition of "force." [111] Moreover, the first law becomes a deductive consequence of this definition. That is if force is equivalent to the product of mass and acceleration, and the force is said to be zero, than the acceleration is zero.

The first "law" then becomes an implicit definition of the absence of impressed forces. This is quite far removed from Berkeley who adduces it as experimental evidence for the "passivity" of bodies (aside from what for him was sufficient *a priori* reasons for such passivity). Although a case can be made for the thesis that Berkeley, like Mach, views mechanics as fundamentally a kinematics, or mathematical description of motion, the case is flawed by Berkeley's failure to in fact dispense with the concept of impressed forces as distinct from and causally related to certain motions.

[111] *Ibid.,* p. 303.

CHAPTER IV

THE PHILOSOPHY OF MATHEMATICS

We have already commented on the formative role the study of mathematics, particulary arithmetic and algebra, played in the development of Berkeley's theory of signs. Mathematics modeled, in his view, some of the crucial aspects of all language. Three elements of this paradigmatic function are, we believe, worth noting. (a) Mathematics reveals more clearly than ordinary language the contingent character of the relation between sign and designatum. This "contingency," of course, is not unique to mathematics; all "conventions of reference" share this trait. In Berkeley's hands, the metaphor of the "ostensive definition," applied to the "causal" relation emphasizes the lack of necessity in this relation. The observed connections in nature are viewed as invariable correlations of types of events established by divine fiat. (b) Mathematics, again, particularly algebra, emphasizes the importance of order, or syntax over reference, to the point, in algebraic "games" [1] where reference plays no role at all and we have a pure formalism. From the syntactical point of view, one can judge the "correctness" (not "truth") of linguistic utterances in term of whether the rules for the combination of "signs" have been correctly followed. Even where reference is important, the importance of syntax emerges in Berkeley's critique of the "Lockean" theory of communication, which appeared to require as a necessary condition for "understanding" language, that each non-logical expression be simultaneously associated with its referent. Berkeley correctly points out, using arithmetic and algebra as examples, that we often use language successfully without paying attention to the referents for individual "signs." We can do this because language involves not only conventions of reference, but conventions of syntax, or rules for combining simple expressions

---

[1] "De Ludo Algebraico," in *Berkeley, Works* ed. Luce, Vol. 4, *op. cit.*, pp. 214-230. (English translation in Rev. G. N. Wright (ed.) *The Works of George Berkeley* (London: 1843) vol. 2))

into more complex ones. We are often in the process of generating "signs" (as in adding a column of figures) where no attention to the referents for individual elements is required.

(c) Berkeley has used examples from mathematics, particularly geometry, to illustrate his contention that linguistic terms gain "generality," not because they refer to some unique ideational content ("abstract general ideas"), but because they have "divided reference" as the perceptual triangle gains generality by being made to stand for any individual triangle.

A general proposition, however, is not necessarily a true one, just as a syntactically correct proposition is not necessarily true. On the other hand, mathematical propositions are said to be apodictic, and have universal validity, appelations Berkeley would generally agree with. The problem then is not merely what mathematics reveals about the general structure of language, but rather, what is the specific difference that distinguishes the "general" propositions of mathematics from those of other sciences, e.g., mechanics.

In attempting to answer this question in the context of Berkeley's writings, we will be concerned in the main with a subsidiary question: what (if anything) are the referents ("objects") for mathematical expressions?

We will deal first with Berkeley's discussion of arithmetic, to be followed by an analysis of his conception of geometry, and finally we will deal with his critique of analysis.

THE PHILOSOPHY OF ARITHMETIC[2]

It is easier to express what Berkeley believed is not, rather than is, the object of arithmetic propositions. Number terms in a language (words or numerals), do not, he says, refer to any "abstract, intellectual, general idea of number" (Alciphron VII 12). There is, then, no "idea" of duality referred to by the numeral "2." We take Berkeley here to be merely extending his critique of the claimed existence of universals *in mente*; there is no ideational content uniquely referred to by the numeral "2," just as there is none for the general terms "red" or "triangle."

Arithmetic has been thought to have for its object abstract ideas of number of which to understand the properties and mutual habitudes is supposed to mean part of speculative knowledge. (*Principles* 119)

---

[2] Our presentation has been influenced by the excellent discussion in, Stephen Korner, *The Philosophy of Mathematics* (London: Hutchinson and Co. Ltd., 1960, Reprinted as a Harper Torchbook; New York: Harper and Row; 1962)

Number terms are significant in the same sense that general classificatory terms (e.g., "man," "triangle") are significant; they have divided reference. Just as "man" can refer to any individual man, "2" would refer to any couple. We can take Berkeley to be saying that number terms refer to properties of collections whereas general classificatory terms refer to properties of individual objects.

The concept of a "collection," however, is never really explicated by Berkeley. Apprehending something as a "collection" cannot be ascribed to a particular sense; at least not in the same way that I can ascribe apprehending that something is "red" to vision, and that something is "sour" to taste. As Frege has pointed out, I cannot read off, as it were, from the world, "numbers" as I can "colors." [3] The concept of a "collection" presupposes the concept of the "unit" or that which is collected or classified. Determination of the unit is not required for the recognition of 'color," although it is for the recognition of "number."

In fairness to Berkeley, it should be pointed out that he recognizes that the numbering of a collection is not a passive apprehension of a property of what is "given" in experience, but presupposes an irreducible conventional element (or decisional element) which is the act of constructing the unit. For example, in the *Essay on Vision*, discussing the question whether numerical properties at least are invariant over sight and touch, Berkeley remarks:

But, for a fuller illustration of this matter, it ought to be considered that number (however some may reckon it among the [4] primary qualities) is noting fixed and settled, really existing in things themselves. It is entirely the creature of the mind, considering either an idea by itself or any combination of ideas to which it gives one name, and so makes it pass for a unit. According as the mind variously combines its ideas, the unit varies; and as the unit, so the number, which is only a collection of units, also varies. We call a window "one," a chimney "one"; and yet a house, in which there

[3] Gottlob Frege, *The Foundations of Arithmetic*, (trans. J. L. Austin, (New York: Harper and Brothers, 1960) p. 29.
"It marks, therefore, an important difference between colour and Number, that a colour such as blue belongs to a surface independently of any choice of ours . . . The Number 1, on the other hand, or 100, or any other Number, cannot be said to belong to the pile of playing cards in its own right, but at most belong to it in view of the way in which we have chosed to regard it; and even then not in such a way that we can simply assign the Number to it as a predicate." Frege quotes with approval (p. 33) the passage from Berkeley's *Essay on Vision*, where the conventionality in the construction of the "unit" is emphasized. (*Essay* 109) ; we quote the full section in the text – p. 247-248.) Without mentioning Berkeley by name, Frege points to the mistake in concluding that because the "unit" is conventional, number is somehow "psychological" or mind-dependent. This is exactly the mistake Berkeley makes.
[4] On "number" as a primary quality; Locke, *Essay Concerning Human Understanding*, Book II, Ch. VIII.

are many windows and many chimneys, has an equal right to be called "one"; and many houses go to the making of one city ... Whatever, therefore, the mind considers as one, that is a unit. Every combination of ideas is considered as one things by the mind, and in token thereof is marked by one name. Now, this naming and combining together of ideas is perfectly arbitrary, and done by the mind in such sort as experience shows it to be the most convenient – without which our ideas had never been collected into such sundry distinct combinations as they are now. (*Essay* 109)

The passage expresses a number of important and inter-related notions; (1) that "number" is a "collection of units"; (2) what counts as a unit is a matter of convention; and (3) the conventionality of classification – the "mind" is free to link any predicates ("ideas") together under the rubric of one "name." The use of the technical term "idea" here is misleading, since the import of the passage is that one could number thoughts (or spirits) as well as tables or windows. It should be remarked in this regard, however, that even given a decision about the "unit," the "number" of a collection is not an "idea" if we mean by "idea" a "proper" (or "immediate") object of any sense (or combination of such objects).

In the *Principles* Berkeley develops a comparable argument that there are no "ideas of numbers in the abstract denoted by the numerical names and figures." The argument is that since there is no "idea" of "unity or unit in abstract," and since "number" is defined as "a collection of units," there can be no "idea" as referent for any other numerical terms. (*Principles* 120) He refers us to section 13, where the claim is, against Locke,[5] that we have no "idea" of "unity." It is not clear, however, how the concept of "unity," or more precisely, how the question of whether "unity" is an "idea," is related to the thesis of the conventionality of the unit. The latter thesis is the one developed in the preceding section of the *Principles*. We quote it in full:

That number is entirely the creature of the mind, even though the other qualities be allowed to exist without, will be evident to whoever considers that the same thing bears a different denomination of number as the mind views it with different respects. Thus the same extension is one, or three, or thirty six, according as the mind considers it with reference to a yard, a foot, or an inch. Number is so visibly relative and dependent of men's understanding that it is strange to think how anyone should give it an absolute existence without the mind. We say one book, one page, or one line' all these are equally units, though some contain several of the others. And in each instance it is plain the unit relates to some particular combination of ideas arbitrarily put together by the mind. (*Principles* 12)

Against Berkeley, however, we would argue that the thesis of the con-

---

[5] *Ibid.,* Book II, Chapter 16, sec. 1.

ventionality of the unit is unrelated to the issue of the mind dependence ("creature of the mind") of numerical properties. Given a certain convention as to the unit, it is an objective property of collections that they have or have not a certain "number" of units. The case, in this respect, is no different then after having established the convention as to what collocation of properties defines a "robin"; it is a fact that a given perceptual datum is or is not a robin. The conventionality of the unit does not entail the kind of perceptual relativity that Berkeley finds with respect to tastes and colors. The perceptual relativity argument [6] is that since certain qualities are not invariant with respect to *changes* in the perceiving subject, they cannot be attributed as "objective" (by which is meant "mind independent") properties of objects. Regardless of the coherence and validity of this view, it has nothing to do with the thesis of the conventionality of the unit. The thesis of perceptual relativity requires, according to Berkeley, that if a particular taste, for example, is said to be an "objective" property of an object, some other and logically incompatible taste could also be such a property. The absurdity of this conclusion entails the falsity of the thesis that taste is an objective property of objects.

There is nothing incompatible, however, about an object being simultaneously one yard, three feet and thirty-six inches in length; as there would be if we said the object was red and blue in the same place. If, *given the convention*, numeration was shown to be a function of changes in the perceiver, we would have a comparable argument.

In addition, it should be mentioned that the concept of the "unit" is not the same as the concept of the cardinal number "1." The former refers to that which is an element of a given class; the latter is the number of a class that contains one and only one of its elements.

Although, to be consistent with the Berkeleian idiom, the "number" of a collection cannot be considered an "idea," we take Berkeley to be contending that number terms in a language are "signs" for certain properties of classes or collection. It should be remembered, however, that in this view there is no "entity" "twoness" referred to by the numeral "2."

---

[6] The argument for the mind dependence of certain qualities, on the grounds of perceptual relativity is found in *Principles* (14, 15) and *Three Dialogues* (Dialogue 1, Turbayne Ed. pp. 116-117). These arguments are taken essentially from Locke, and are not considered by Berkeley necessary to establishing the mind-dependence of "secondary" qualities. His stronger argument is that it is constitutive of the concept of "idea" or object of sense that "ideas" cannot exist independently of minds. From this point, one could say the "number" of a collection is mind-dependent in the sense that what is being counted are "ideas." However, if we wish to use "mind-dependence" in this sense, it has nothing to do with the conventionality of the unit.

Like "red" the numeral "2" can be used to refer to certain objects (which have to be construed for number terms as collections) that have in fact a certain property, being red, or having two elements.

On the other hand, Berkeley continually suggests that "things" (by which we again have to understand classes or collections) are the mediate objects of arithmetic; its immediate objects are the arithmetic "signs" themselves.

In arithmetic, therefore, we regard not the things, but the signs, which nevertheless are not regarded for their own sake, but because they direct us how to act with relation to things, and dispose rightly of them. (*Principles* 122)

The point is echoed in *Alciphron:* after briefly discussing the genesis of arithmetical notation and the rules of computation, Berkeley comments:

. . . in which theory and operations, the mind is immediately occupied about the signs or notes, by mediation of which it is directed to act about things, or number in the concrete (as logicians call it) without even considering the simple, abstract, intellectual general idea of number. (*Alciphron* VII 12)

In an important sense, this distinction between the immediate and mediate objects of arithmetic is a *practical* one, and should not be hastily construed as a distinction between "pure" and "applied" arithmetic. This practical perspective surfaces clearly in another passage from the *Principles.*

For these signs being known, we can by the operations of arithmetic know the signs of any part of the particular sums signified by them; and, thus computing in signs (because of the connection established betwixt them and the distinct multitude of things whereof one is taken for a unit), we may be able rightly to sum up, divide, and proportion the things themselves that we intend to number. (*Principles* 120)

The clearest meaning of the passage is that in using the algorithms of arithmetic, no attention need be paid to the "things" (empirical collections) referred to by the numerical "signs," including among the latter, "signs" for the various operations, such as addition, subtraction, etc. Arithmetic, again, presents a clear example to Berkeley of the ability to use "signs," without simultaneously focusing on their designata.

The issue raised by the above view concerns not the practicality but the truth of the propositions of arithmetic. If arithmetic, albeit mediately, refers to operations upon empirical collections, can I conclude that such propositions could by false, as for example that "two" objects added (some physical definition of addition would have to be specified) to "three" objects might not give "five" objects? The problem more generally stated is whether the view that arithmetic deals with ultimately actual operations (adding, dividing, etc.) on empirical collections is compatible with the view that the propositions of arithmetic are necessarily true.

There is some evidence, unfortunately scanty, that Berkeley did consider the propositions of arithmetic to be necessarily true (of universal validity) and that this view was connected with his contention that the "immediate" objects of arithmetic were "signs." Some of the early entries in the *Philosophic Commentaries* suggest this association:

Reasoning there may be about things or ideas and actions, but Demonstration can be only verbal . . . (# 804)

The entry, however, does not distinguish arithmetic, for example, from logic, although it does suggest that "demonstration" which we take to mean a logical "proof" is somehow essentially related to the use of signs. In the same vein is the following entry in the *Commentaries*:

The reason why we can demonstrate so well about signs is that they are perfectly arbitrary and in our power, made at our pleasure. (*Philosophic Commentaries* 732)

Warnock takes the passage as evidence that "Berkeley was clearly aware" of the difference between "using arithmetic in 'reckoning,' counting and so on, and doing 'pure' mathematics." [7]

Warnock, however, does not make clear what sense of "pure" mathematics he is referring to. It could mean arithmetic as an axiomatic system (such as 'Peanos') where certain primitive terms have *no* extra-systematic reference, but are contextually defined by the axioms. As definitions, however, we would not call the axioms "true," although we could grant systematic truth to the theorems as logical consequences of the axioms.

There is little evidence that Berkeley viewed arithmetic in this way; not even as a semantically significant system where the "axioms" would have extra-systematic reference, but might be considered "intuitive" or self-evident truths.[8] On the other hand one might consider Berkeley's claim that arithmetic as a "demonstrative" science deals with "signs" and not "things" is a "formalist" position somewhat akin to the modern formalist views represented by David Hilbert.[9] We are not thinking here

[7] Warnock, *op. cit.*, p. 206.

[8] Even in the "Arithmetica and Miscellenea Mathematica," (trans. Wright, *op. cit.*, Vol. 2) there is no such attempt.

[9] David Hilbert, "On the Infinite," trans. Erna Putnam and Gerald J. Massey from Mathematische Annalen (Berlin No. 95, 1925) pp. 160-190. Reprinted in Paul Bena-

so much of the problems of axiomatization or consistency, but the view that the objects of arithmetic are certain perceptual objects, marks on a piece of paper or blackboard.

For Hilbert, however, "proof" is fundamentally tied to perceptual intuition. The series of natural numbers is built up in terms of "strokes," (e.g., "2" $= //$) and the propositions of arithmetic can be determined to be true by a series of finite inspections (proof "ad oculos"). For example, that "$1 + 1 = 2$" can be demonstrated by noting that the production of a single stroke ($/$) and then another, gives us the perceptual object$//$, which is what is meant by "2." The proposition "$3 > 2$," corresponds to the recognition that the configuration $///$ requires one more stroke than $//$.

Although Berkeley discusses the use of a stroke function in counting, there is nothing of this intuitionist strain in his writings with respect to proof. On the other hand, focusing merely on the algorithms of arithmetic, one could imagine arithmetic as a "formalism" from his point of view. Given the natural number numerals, "0" through "9," they can be combined in addition and multiplication tables (introducing the signs "$+$" and "x"). The individual numerals and the numeral combinations in the tables are perceptual objects. The numeral combinations in the tables would function both as rules for combining numerals and as a check on the correctness of any given combination. "Proof" here of the correctness of a complex arithmetical proposition would involve a finite series of inspections, which involve checking combinations against the table. The required *rules* for "associativity," "commutativity," etc., the rules for the use of relations like "equals," or "greater than," could themselves be considered perceptual objects, with the important difference that individual elements would have the status of variables as opposed to constants.

There is no need to flesh out the suggestion in more detail. The point is that arithmetic can be done purely in terms of the combination according to rules, of perceptual objects, where no claim is made concerning any extra arithmetic reference for these objects. At least this is how we take Berkeley to understand arithmetic as a formalism.

There is a distinct difficulty with this view, comparable to the one raised by Körner against the formalism of Hilbert.[10] If the "immediate" objects of arithmetic (or, if we wish, the objects of "pure" arithmetic)

cerraf and Hilary Putnam ed., *Philosophy of Mathematics,* (Englewood Cliffs: Prentice Hall, 1964) pp. 134-151. See also S. Korner, *op. cit.,* Ch. IV and V.

[10] Korner, *op. cit.,* p. 103.

are perceptual objects – objects with particular spatio-temporal locus – what tells us that a given "formula" or sequence of signs is sufficiently like the paradigmatic one in the table to be judged "correct"? Certain parts of the perceptual configurations are considered unimportant in this respect; e.g., the size of the individual numerals, or the spacing between the elements in a particular "formula." What, in fact, are the essential elements of the perceptual object, "1 + 1 = 2," that offer a basis of comparsion with other perceptual objects?

A convinced "nominalist" might rejoin that we have merely raised once more the issue of how one judges whether a particular item of perceptual experience (e.g., a "color") is a member of a particular class. There is no more of a problem here than in deciding whether a particular item of experience is to be called an "elephant." The "nominalist" would admit *this* interesting difference, that whereas the term "elephant" or a pictorial elephant schema is not a member of the class of elephants, the paradigmatic configuration, "1 + 1 = 2," is a member of the class of which all other members are its copies, much as the "standard meter" is a member of the class of objects which are said to be a "meter" in length.

We might conclude from Berkeley's very brief remarks on arithmetic, that doing "pure" arithmetic is merely focusing on arithmetic terms or "signs" not in their function as signs but in their status as perceptual objects; where certain configurations of "signs" have a *directive* function, that is, they tell us how to combine certain perceptual configurations (the "numerals" for example). This directive function, which constitutes a configuration as a "rule," is, of course, not an element of its status as a perceptual object but a status given to it by the arithmetician. It is in this sense that we might understand the passage in the *Commentaries,* quoted by Warnock, to the effect that our ability to "demonstrate" with signs is a function of the fact that they are "arbitrary" and "in our power."

If our view of what Berkeley would consider "pure" arithmetic is correct, we would have to modify our contention that he would take the propositions of arithmetic to be necessarily true. Rules are not propositions, in the sense that a convention as to how perceptual objects should be combined is not true or false. It is important, in this regard, to remember that when Berkeley speaks of the "object" of arithmetic, he means its "subject matter," and not the referents for individual arithmetic terms. Viewed as a formalism in the above sense, arithmetic would have a subject matter, but no referents for the individual signs.

The question of "necessary truth" would have to apply to applied arithmetic – arithmetic as dealing with "things," or empirical collections. However, there is no evidence that Berkeley entertained questions such as, if I add two apples to three apples (where "two" and "three" are identified by counting) the result must be five apples (again "five" would be identified by a process of counting). There is no evidence that he would question the necessary truth of such propositions; yet, as well, we have no indication of the kind of analysis he would make of propositions both "necessarily true," and applicable to "things."

PHILOSOPHY OF GEOMETRY

In an entry in the *Philosophic Commentaries,* Berkeley contrasts geometry, which he calls "mixed mathematics with arithmetic and algebra, which he calls "purely nominal." (*Philosophic Commentaries* 770) Geometry, he suggests here, is the application of arithmetic and algebra to "points." The view is amplified in the *Essay on Vision* where it is claimed that the "object" of geometry (subject matter) is "tangible extension."

Berkeley insists that there is no "idea" of extension simpliciter, which is the subject-matter of geometry. Although this follows straightforwardly from the definition of "idea" as "object of sense," it can be taken also as a rejection of the claim that Euclidean geometry articulates the properties on a "entity" called "space" conceived of as being distinct from and the container of physical events and processes. The historical antecedent for Berkeley's views is more appropriately Aristotle and the view that geometry articulates the properties of the boundaries of objects which are conceptually but not physically separable from the objects themselves.

Berkeley also suggests in the *Essay* that not only is it perceptual extension which is the subject-matter of geometry, but that it is properly "tangible extension" which is the object of geometry; the visual figures in geometrical diagrams merely being signs for tangible designata, which are the subjects of predication in the propositions of geometry.

This claim appears to be based on the following considerations: (1) Geometry, Berkeley holds, is the science of magnitude, or the science by which we measure extensive magnitudes; such mensuration is a precondition for having mathematically formulated scientific laws. For such mensuration to be useful, the determination of length must be invariant not only across subjects (inter-subjective agreement) but with respect to the visual perspective of a single subject; more precisely invariant with respect to the distance of the object to be measured from the subject.

"Visual magnitude," as Berkeley defines it, does not meet this latter condition, and therefore the invariance we seek must be predicated of tangible magnitude. We have already discussed this, noting that nowhere does Berkeley show how the requisite invariance is uniquely related to the data of touch. (2) The determination of magnitude is based on the establishment of a coincidence or congruence relation between a standard unit and the magnitude to be measured; this Berkeley calls the "application of tangible extension." What this appears to mean is that the physical boundaries of objects are shown to be coincident; so that the term "tangible" refers to physical boundary. Again, however, there is no particular relation between the operation of measurement and the data of touch; the determination of coincidence itself is often, if not invariably made by sight.

(3) Berkeley argues that geometry as the science of solid bodies, and even planes; to the extent that a "plane" is viewed as the section of a solid; requires a sense or apprehension of space, (tri-dimensionality) and this "sense" could only be developed in a being with tactile sensations. He entertains, in the *Essay* the interesting question, whether a being endowed only with the "visive faculty," could develop geometry. (*Essay* 153-155) Since he would have no idea of "distance," such a being, Berkeley claims, could have no conception of solids, and therefore would lack that part of geometry dealing with the "mensuration of solids." (*Essay* 154) Nor, for comparable reasons, would he have any idea of convex or concave surfaces. In addition, such a being could have no idea of how geometers "describe" lines and circles. The apprehension of the "spatial field" as tri-dimensional is required, according to this view, in order to construct the geometric figures used in demonstrations. Equally important, the concept of "superposition" required to establish the fundamental theorems about the congruence of figures requires, Berkeley claims, the "idea of external space."

Assuming that the above account were correct in terms of the genesis of geometry, it is not clear how it would establish that the object of geometry is tangible magnitude. The central ambiguity in the concept of "tangible magnitude," remains: whether it refers to the boundaries of physical objects assumed to remain invariant over transport, or has some special reference to tactile sense data. There is also an ambiguity as to the role Berkeley gives to the geometric "constructions" used in demonstrations. We will return to this question when we consider Berkeley treatment of geometric "demonstration" or proof, but the central issue can be mentioned now: are the constructions merely aids in the demonstration, without any essential relation to it, or is the "truth" of the Euclidean theorems demonstrated *by* the possibility of the construction? [11] It would be absurd to fault Berkeley for not developing geometry as a formal system comparable to Hilberts' which would eschew any visual representation. Yet as far back as Berkeley's early critic Bailey, the need for the concept of superposition in the demonstration of congruence had been questioned.[12] And as modern developments of geometry make clear, the imaginative conception of "placing one plane or angle on another" (*Essay* 155), is not required in the demonstration of congruence.

In the *Principles, De Motu,* and the *Analyst,* Berkeley appears to drop this emphasis on tangible extension as the subject-matter of geometry, and in our own subsequent discussions we will assume that extensive magnitudes, figures, etc., apprehended visually are in fact the distinct subject-matter of geometry, since the distinction, even if maintained, plays no role in these discussions.

Sensible extension (visual or tangible) is, according to Berkeley, composed of "points" or "minima." We have discussed to some extent the grounds for Berkeley's contention that finite perceptual extension is not infinitely divisible; and it is interesting to consider the consequences when this thesis is wedded to the view that perceptual extension is the object of geometry.

It is difficult not to agree with Warnock's contention that the view that perceptual extension is composed of "minima," conjoined with the belief that such extension is the subject-matter of geometry would "wreak havoc" in this discipline.[13] Finite segments of an *even* number of minima could not be bisected, for example (this as against Warnock, who suggests the problem would be with segments of an *odd* number. We would have to assume, however, that the width of the dividing segment was at least one minimum). The Pythagorean Theorem, as Berkeley admits himself, would have limited validity. Fundamentally geometry as the science of extension would have to deny the property of "denseness" for finite seg-

---

[11] Berkeley's views are also somewhat akin to those expressed by Newton in the Preface to the 1st edition of the *Principia* (Newton, *Principia, op. cit.,* p. 1). "To describe right lines and circles are problems, but not geometric problems is required from mechanics, and by geometry the use of them, when so solved is shown; and it is the glory of geometry that from those few principles fetched from without, it is able to produce so many things. Therefore geometry is founded in mechanical pratice, and is nothing but that part of universal mechanics which accurately proposes the art of measuring." We will discuss how geometric constructions are related to the question of "demonstration" in geometry. The fundamental problem seems to be this: how does one guarantee that the *actually* constructed figures have the properties attributed to them by the Euclidean axioms and theorems?

[12] Bailey, *op. cit.*

[13] Warnock, *op. cit.,* p. 209.

ments, a property entailed by the alleged universality of the theorem on bisection. It is not merely that the square's diagonal could not be incommensurate with the side; strictly speaking, no non-integral value for any length would be allowable, unless the unit of measurement was more than one minimum; in which case the lower bound for a fractional value would be $1/n$ (where n was the number of minima in the unit).

A defender of Berkeley might point out that there is clearly a physical limit to the actual bisection of a line segment, and a perceptual limit to how finely one can "measure" any given segment. Therefore there is a clear limitation in the empirical domain of the applicability of the Euclidean theorems.

Such a defense is subject, however, to a number of objections. In the first place, a consequence of Berkeley's view (as interpreted above) would appear to be that indirect measurement making use of the Euclidean axioms would be disallowed. We might be faced with the paradoxical problem of allowing the axioms to apply to the sides of a square (the "sides" would be considered "straight" in the Euclidean sense) while denying their applicability to the diagonal. If the axioms are said to be true of a given empirical domain, it follows logically that the theorems are true of that domain. Conversely, if we deny the truth of the theorems, the truth of the axioms must be denied. Berkeley's empirical reformulation of geometry, based on the alleged facts of spatial perception, would require a fundamental reformulation of the axioms; and there is no evidence that Berkeley has attempted this.

This first objection leads to a more general consideration. If one claims that geometry as a method of measurement has a limited domain of applicability because of certain perceptual considerations, a distinction is presupposed between "pure" and "applied" geometry. There *are* passages throughout Berkeley's works (not just in the later works) that suggest a conception of "pure" geometry; understanding by this that its primitive terms have no empirical referents. For example, we may consider the previously quoted passage from the *Essay on Vision:*

The truth of this assertion [that distance is not immediately perceived] will be yet further evident to anyone that considers those lines and angles have no real existence in nature, being only a hypothesis framed by the mathematicians, and by them introduced into optics that they might treat of that science in a geometrical way. (*Essay* 14)

A comparable view is expressed in *De Motu:*

And just as geometers for the sake of their art make use of many devices which they themselves cannot describe nor find in the nature of things, even so the mechanician makes use of certain abstract and general terms, imagining in bodies force, action, attraction, solicitation, etc. which are of first utility for theories and formulations, as also for computations about motion, even if in the truth of things, and in bodies actually existing, they would be looked for in vain, just like the geometer's fictions made by mathematical abstraction. (*De Motu* 39)

If the "lines" and "angles" of Euclidean geometry are considered "fictions" made by "abstraction," in what sense is this to be understood? One normally things of the notion of 'idealization"; that the objects of "pure" geometry are not perceptual objects, but idealizations of these representing the perfect satisfaction of the axioms, e.g., extensionless points, lines without breadth, etc. The claim is that such "objects" do not exist (are not "ideas" in the Berkeleian idiom) but are useful in the computation of magnitudes.

And just as by the application of geometrical theorems, the size of particular bodies are measured, so also by the application of the universal theorems of mechanics, the movements of any parts of the mundane system, and the phenomena thereon depending, become known and are determined. (*De Motu* 38)

It is, however, precisely in this "application" of geometry to measurement that the difficulty previously mentioned arises. In order to apply geometry, some empirical domain must be assumed to satisfy the axioms; e.g., "light rays" or the edges of rules will be considered "straight." We cannot say, then, without contradiction, that the theorems do not apply to this domain. Yet this is what we apparently are forced to do, if in fact the theorems impute magnitudes to objects that could not, according to the doctrine of sensible minima, be possible magnitudes.

The above remarks lead to a more generalized objection to the defender of Berkeley, who might claim that the doctrine of sensible minima correctly recognizes the empirical limitations built into any use of chalk lines, pencil points, rules, light rays, etc., as geometrical objects. The seriousness of this objection depends on how seriously we consider Berkeley to be offering, in the doctrine of sensible minima, a method for the metrization of extensive magnitudes to be *used* in geometrical optics or mechanics. The suggestion would be that perceptual extension has an *intrinsic* metric; that is each individual linear segment can be given a number corresponding to the number of minima it contains, no comparison with any other segment being required. This would be contrasted with a relational or extrinsic metric; that is, magnitudes are assigned to lengths (other than the conventionally chosen standard) in terms of the latters' relation to some other length. This can be done through direct comparison with the standard, or through the application of the Euclidean theorems.

Even admitting that in certain distinct types of spatial perception, for example, Hume's image of a spot diminishing to zero, we intuitively sense there is a "jump" from a last finite magnitude to disappearance,[14] it is difficult to see how this intuitively grasped "minimum" could form the foundation of any metrization of extensive magnitudes. As previously mentioned, we are not aware of either visual or tangible extension as aggregates of "points" or "minima." It is not even clear that it is meaningful to speak of "minima" being equal to each other. If they are equal it is certainly not demonstrated through any actual comparison, but is an equality that follows merely from the meaning of the term "minimum." The closest, perhaps, that Berkeley came to establishing a metric based on "visual minima," was in thinking of the latter in terms of the proportion a certain datum occupied in the visual field. Here, too, we have resorted to a comparison, and are no longer thinking of a metric intrinsic to a particular sense. Moreover, Berkeley often tended to think of the "proportionality" in terms of the proportion the "image" of the datum occupied on the retina. Here, however, the metric – that is the magnitude of the retina – the magnitude of the particular image, and the ratio of the two are not established by vision itself.

Regardless of the success of the program to establish an intrinsic metric based on aggregates of minima, Berkeley is raising an important question, which can be formulated as follows: on what basis do we decide that physical bodies qua extended magnitudes, or the "space" in which they move (their "paths") have certain properties identical with the properties of number systems we might use to study them? How would one decide whether "physical space" was discrete (consisted of points in one-to-one correspondence with the whole numbers), or "dense" (in one-to-one correspondence with the rationals), or "continuous" (in one-to-one correspondence with the real numbers)? An investigation of this question is beyond the scope of this essay.[15] However, if there were an intrinsic

[14] See the interesting discussion in Adolf Grunbaum, *Modern Science and Zeno's Paradoxes*, (Middletown: Wesleyan University Press, 1967) Chapter II., sec. 1 and 2. Grunbaum argues that with respect to the perception of temporal segments, there are perceptual minima; that is the awareness of time is an awareness of a discrete sequence of "nows," each of which must be said to have some minimum but actual duration. He argues however, that our perception of space is not in any comparable sense an awareness of aggregates of spatial minima. The *disappearance* of the "spot" in Hume's example, might then more validly demonstrate the discreteness of temporal perception.

[15] A good discussion of this problem is found in, A. Grunbaum, *Geometry and Chronometery in Philosophical Perspective*, (Minneapolis: University of Minnesota Press, 1962).

metric to perceptual space, based on the summation of aggregates of minima, we would still have the problem of whether the "discreteness" of perceptual space disallows the "denseness" of physical space. For Berkeley, presumably, the problem does not arise, since from the point of view of immaterialism there is no distinction between perceptual and physical extension; there are no unperceived extensive magnitudes.

Here again, however, we would have a conflict between the consequences of the application of geometry and consequences of the doctrine of sensible minima *used* as the basis of the metrization of space. For example, take three "empirical" lines (chalk lines, pencil lines, light rays, etc. judged to be "straight" in the Euclidean sense); one, a base line (K) units; and two lines making equal acute angles with the base at its end points. At a certain distance from the base, the "distance" between the two lines will be unperceivable, and therefore zero according to the doctrine of sensible minima; or more precisely according to the conjoined doctrines of sensible minima and immaterialism. Yet it would be a consequence of calling them "straight" that there would be some distance between them which is a function of (K), the length of the sides and the angles made with the base.

It should be remembered that when Berkeley speaks of the "object" of geometry, this refers in the main to its field of application. He denies both that this "field" is extension simpliciter, which means extension as an existing entity (absolute space); or that it is certain "general abstract ideas," e.g., generic triangles, circles, etc. which for Berkeley could not in principle *be* objects of sense of "ideas."

The passages in the *Principles,* for example, where the question of the "object of geometry" is explicitly discussed (Sec. 123-128), are concerned with establishing that finite perceptual extension, held to be the subject-matter of geometry, is not infinitely divisible.

Berkeley offers a rationale for his belief that men have been misled into considering finite line segments to be infinitely divisible. This "rationale" although, in our view, fundamentally mistaken, is worth some consideration. Reduced to its essentials, the error, according to Berkeley, of attributing infinite divisibility to a finite segment stems from a confusion between sign and designatum, particularly, a confusion between diagrams in geometrical demonstrations and what they signify. Berkeley's argument can be outlined as follows:

1. The actual magnitude of a given line *in* the demonstration figure is irrelevant to the demonstration.

2. The universality of the demonstration lies in the fact that the segment in the diagram "represents" or "stands for" all possible lines.

3. Therefore when we speak of a line as containing "10,000 parts" or "innumerable parts," we mean no more than that it signifies all "lines" "larger than itself" which can in fact be divided into 10,000 parts, or an "infinitely large" line (Sec. 128) which could in principle be infinitely subdivided.

However, the confusion Berkeley speaks of appears to be one of his own invention. In labeling the legs of a right triangle 3/10,000 and 4/10,000 of an inch respectively, and concluding that the hypotenuse is 5/10,000 of an inch, no claim is necessarily made concerning the actual magnitudes of the lines in the diagram, although there is the claim that *if* the legs had such magnitudes, the hypotenuse would have the deduced magnitude. The error of confusing magnitude labels with actual magnitudes is not one we can imagine any geometer falling into; hence Berkeley's argument here is a straw man. Berkeley contends (Sec. 127) that although for practical purposes we can ignore differences of 1/10,000 of an inch, we cannot ignore differences of 1/10,0000 of a mile. Even if true, it is not clear how this fact would compel us as he says (Sec. 128) to speak "of lines described on paper as though they contained parts which they really do not," unless we were suffering from the illusion that magnitude labels in demonstrations were magnitude *designations* of the segments in the diagram. But this illusion, again, is one that no geometer suffers from.

Berkeley is correct in holding that the labels we put on lines in the diagram do not necessarily designate the magnitudes of these lines; he is wrong in suggesting that the claim of infinite divisibility for the perceptual lines in the diagram stems from ignoring this distinction.

Berkeley does refer us (*Principles* 126) back to the Introduction to the *Principles,* where the issue of geometrical demonstration is discussed. The problem of demonstration is particularly important for him, for he hopes to put to rest the belief that the universality of geometric propositions, a universality allegedly grounded in their being logically "demonstrated," implies they are not about particular perceptual figures, but about abstract geometrical entities, triangles, circles, etc. which are in fact the subject of demonstrations.

Berkeley's response is, on the surface, a quite simple application of his theory of signs. Geometrical demonstrations *are* about perceptual objects, but not uniquely about the perceptual object that figures in the demonstration.

Suppose a geometrician is demonstrating the method of cutting a line in two equal parts. He draws, for instance, a black line of an inch in length: this; which in itself is a particular line, is nevertheless with regard to its signification general, since as it is there used, it represents all particular lines whatsoever; for that which is demonstrated of it is demonstrated of all lines, or in other words of a line in general. (*Principles* – Intro. 12)

Put formally: when I demonstrate that figure (F) has the property (P), since (F) stands for or "represents" all other figures of the same type, I have demonstrated that every (F) has the property (P).

An immediate problem is on what grounds do I identify a perceptual configuration as an example of (F)? If some perceptual identification is meant, it would seem quite possible that an (F) so identified would not have the property (P). We are assuming that having the property (P) or any property that entails having (P) is not part of the criteria for identification. How are we to know, for example, that because one perceptually identified line segment can be bisected, any other can. Merely saying that the bisected segment "stands for" all others, hardly insures that the bisection can be performed on any other segment.

The problem, of course, is making sense of what it means to demonstrate something of a particular perceptual figure and yet insure that what is demonstrated is not true of that figure alone. Berkeley, in the following important passage, recognizes the problem, and proposes a solution:

But here it will be demanded how we can know any proposition to be true of all particular triangles, except we have first seen it demonstrated of the abstract idea of a triangle which equally agrees to all? For, because a property may be demonstrated to agree to some one particular triangle, it will not thence follow that it equally belongs to any other triangle which in all respects is not the same with it. For example, having demonstrated that the three angles of an isosceles rectangular triangle are equal to two right ones, I cannot therefore conclude this affection agrees to all other triangles which have neither a right angle nor two equal sides. It seems therefore that, to be certain this proposition is universally true, we must either make a particular demonstration for every particular triangle, which is impossible, or once for all demonstrate it of the abstract idea of a triangle in which all the particulars do indifferently partake and by which they are all equally represented. (*Principles* 16 – Intro.)

After this rather admirable statement of the difficulty, Berkeley suggests the following way out:

... though the idea I have in view whilst I make the demonstration be, for instance, that of an isosceles rectangular triangle whose sides are of a determinate length, I may nevertheless be certain it extends to all other rectilinear triangles, of what sort or bigness soever. And that because neither the right angle, nor the equality, nor determinate lengths of the sides are at all concerned in the demonstration. It is true the diagram I have in view includes all these particulars, but then there is not the least mention made of them in the proof of the proposition. It is not said that the three angles are equal to two right ones, because one of them is a right angle, or because the sides comprehending it are of the same length. Which sufficiently shows that the right angle might have been oblique, and the sides unequal, and for all that the demonstration have held good. And for this reason it is that I conclude that to be true of any oblique-angular or scalenon which I had demonstrated of a particular right-angled equicrural triangle and not because I demonstrated the proposition of the abstract idea of a triangle (*Principles* 16 – Intro.)

Although Berkeley is not at all clear about the nature of "demonstration," we might take the above passage to be making the following claim: to demonstrate that (F) is (P) is to show that it is a logical consequence of (x) being an (F) that (x) has the property (P). With respect to the above example, it is a logical consequence of the Euclidean axioms assumed to be universally valid) and the fact that (x) is an (F), (a figure bounded by three Euclidean "straight" lines) that (x) has the property (P) (the sum of its interior angles is equal to two right angles).

This interpretation, however, appears to ignore the following important consideration: although Berkeley speaks of focusing, in the demonstration, merely on the generic properties of the figure ("considering a figure merely as triangular" (Sec. 16)), he clearly suggests that the demonstration *is* somehow concerned with the particular perceptual figure in the diagram, presupposing that this figure has been identified independently of the demonstration as an "isosceles rectangular triangle." What assures us that this particular figure ("idea") is in fact such a triangle? If we answer that it is perceptual (visual) identification that is the criterion, we are faced with the previously raised problem: if the demonstration is about this particular figure, how can we be assured that what is shown to be true of it is true of any other figure perceptually identified as the same? What assures us that the spatial position of our figure is not crucial to the demonstration? The mere fact that this position is not "mentioned" in the demonstration cannot assure us that it is not in some essential way made use of.

The above remarks lead, in fact, to the following reformulation of the question concerning the "object(s)" of the demonstration. In what sense, if any, can we understand the "demonstration" even to be about the particular figure in the diagram? The problem was clearly recognized by Husserl in his critique of Berkeley's theory of "abstraction" as "selec-

tive attention" to allegedly universal elements in a particular configuration. We quote from Husserl's English expositor, Marvin Farber.[16]

He [Berkeley] spoke as though a geometrical proof referred to an ink triangle
on the paper or a chalk triangle on the board, and as though in general
thinking the single objects accidentally occuring to us were the objects of
thought instead of mere supports of our thought intention. No geometrical
proposition holds for that which is drawn in the physical sense, because in
reality there can never be a straight or geometrical figure. In no act of
thought does the mathematician mean the drawing; he means rather 'a
straight line in general.'

We would take issue, however, with the claim that "in reality there
can never be a straight or geometrical figure." It is certainly possible to
consider the Euclidean axioms as empirical generalizations about the
properties of physical objects such as light rays, "straight" edges, etc. The
theorems, although deduced ("demonstrated"), would inherit the empirical or synthetic character of the axioms. From this point of view, the
claim that there is a perfectly straight line might, like any empirical proposition, turn out to be true. The conception of physical objects as „approximations" to ideal geometrical figures presupposes that such "ideal"
objects are the referents for the geometrical terms. Without this presupposition there seems no reason to claim that the Euclidean axioms may
not be truly predicated of perceptual (or physical) objects.

Can we take Berkeley to be espousing such a straightforwardly empirical view of Euclidean geometry? After all, Berkeley continually claims
that the subject matter of geometry is perceptual extension. There are,
however, distinct difficulties with such a proposal. In the first place, it is
still unclear in what sense a geometrical demonstration is about the perceptual figure in the diagram. This question itself can be divided into two
subquestions: (a) on what basis do I identify the perceptual object as a
certain "figure" (e.g., a triangle); and (b) in what sense do I demonstrate *of* this figure that it has a certain property (e.g., the sum of its
interior angles is equal to two right angles)? With respect to both (a) and
(b), one might suggest that Berkeley accepted what Helmholtz aptly
called "Euclid's method of constructive intuition," [17] where a figure being
of a certain type (e.g., a triangle), and having a certain property (e.g., the

[16] Marvin Farber, *The Foundations of Phenomenology*, (Albany: State University
of New York Press, 1943) p. 263.

[17] Hermann Von Helmholtz, "On the Origin and Significance of Geometrical
Axioms," (originally given as a lecture in Heidelberg in 1870, Reprinted in James R.
Newman ed. *The World of Mathematics*, (New York: Simon and Schuster, 1956)
Vol. 1, pp. 647-671.

sum of its interior angles is equal to two right angles), is said to be exhibited by actual constructive procedures, utilizing for example, straight edge and compass. The claim might be that the construction guarantees that the figure is a triangle, by which is meant a figure bounded by Euclidean "straight lines"; that is lines are assumed to satisfy the Euclidean axioms. Moreover the possibility of the construction of a *unique* parallel (to the base) through the vertex, (a construction which "embodies" the parallel postulate) is essential in showing that the sum of the triangle's interior angles is equal to two right angles.

"Demonstration" in this sense means essentially construction of the object; viewing this as either constructing a certain figure, or ultilizing constructions to reveal a property possessed by the figure. However, this sense of "demonstration" must be distinguished from logical deduction. There can be no logical guarantee that the actually constructed lines are "straight" in the sense of satisfying the Euclidean axioms or that the constructed "parallel" satisfied the "fifth" postulate. In addition, Berkeley's brief references to "demonstration" lead us to surmise that he had in mind the process of logical deduction and not the process of construction outlined above.

The second difficulty with the suggestion that Berkeley views geometry as a straightforwardly empirical discipline, concerns the claim of universal validity of its propositions. Even if it is accepted than an actual construction can guarantee that a particular figure has a property, there is no assurance that the construction can be repeated. Helmholtz comments to the effect that the use of constructions to support the claim of universal validity smuggles in universal elements which cannot a priori be guaranteed by any particular construction. We would have to assume, for example, that the spatial position of the figure is irrelevant to the possibility of the construction.[18]

There is a way of interpreting Berkeley's remark that in "demonstrations" we can "consider a figure merely as triangular" (*Principles* 16 – Intro.) an interpretation, however, that appears incompatible with the view that perceptual extension is the object of geometry. We can treat the diagrams as aids to, but not essential in the demonstration. The "demonstration" then has a purely hypothetical character; if (x) is an Euclidean triangle (bounded by "straight lines" – those assumed to satisfy the Euclidean axioms), then the sum of its interior angles must equal two right angles. From this perspective "paying attention" to the

---

[18] *Ibid.*

figure generically, or merely as a triangle, has the sense of articulating the implication of something being a Euclidean triangle. Demonstrations in this sense have no more existential import than articulating the meaning of "bachelor" implies anything about the existence of bachelors. Berkeley himself, after speaking of considering a figure "merely as triangular," comments:

In like manner we may consider Peter so far forth as man, or so far forth as animal, without framing the forementioned abstract idea either of man or of animal, inasmuch as all that is perceived is not considered. (*Principles* – Intro. 16)

The simplest interpretation of this is that we can draw the consequences, *if* Peter is a man, or *if* Peter is an animal. The process of demonstration involves no claim that "Peter" is either one or the other.

This view of "demonstration," which carefully distinguishes explicating a concept from claiming existential import, can be accepted as Berkeley's meaning, however, only at the expense of what Berkeleian texts we have concerning geometry. His view in the *Commentaries,* that only arithmetic and algebra are purely "nominal" sciences is, in our view, maintained in the *Principles.* "Paying attention" to the allegedly universal elements in the diagram is not for him merely explicating a concept (which would make geometry a science "merely nominal"), but is somehow connected with the particular figure in the diagram, although its "particularity" is discounted. In our judgment, Berkeley wishes to maintain both that the demonstration is about the particular figure in the diagram *and* that the demonstrated proposition has universal validity.

From this point of view, we can partially agree with Husserl's criticism: I cannot necessarily presuppose that the figure before me has a particular geometric property. And without this presupposition the view that "demonstration" truly attributes a property to the figure in the diagram cannot be maintained. On the other hand, if I judge, through some empirical means, for example, a construction, that the figure before me is a Euclidean triangle, I can articulate the logical consequences of its being such a figure, as, *after* judging that Peter is a man, I can conclude he is an animal. However, this logical "articulation" cannot insure that the original construction was accurate, or that any other constructed figure will be an Euclidean triangle. In this sense the demonstration lacks universal validity, if by the latter we mean that the property demonstrated is necessarily predicated of any perceptually identified figure, including the figure in the diagram.

We would conclude that maintaining the "necessary" truth of the propositions of geometry must be at the expense of the claim that we are demonstrating that certain perceptual objects – those in the diagram, and those they are said to "stand for" – have particular geometric properties. In the *Principles,* at least, Berkeley does not make good what Wild believes to be his ultimate aim, to develop a view of mathematics that would allow its propositions to be both necessarily true, and not merely "nominal," that is, merely "conversant about signs." [19]

It must be remembered that Berkeley's discussion of geometry in the Introduction to the *Principles* occurs in the context of a discussion of the significance of general terms. The geometrician's *act* of letting a particular drawn "line" stand for any other line, is viewed as analogous to a linguistic decision to allow the term "line" to refer to any perceptually identified line, or the term "red" to refer to any red object. Both the drawn line of the geometrician and the term "line" are "signs."

Unfortunately, and perhaps because he is focusing on the problem of *general significance* and not *necessary truth,* Berkeley blurs important distinctions between geometric diagrams and general terms in language. The term "line" is not necessarily a member of the class of lines (although we could use, if we wish, an actual line as substitute for the term "line"). The geometrician's line is, on the other hand, *for Berkeley,* necessarily a member of the class whose members it distributively refers to, because he believes things done to it (through constructions, for example) and demonstrated of it, apply to all its designata. The geometrician's line is given a double and untenable function: it is said to be the subject of the demonstration (the property is demonstrated *of it*), and because it functions (by decision) as a general term, the property demonstrated of it is said to be true of all its designata. Clearly, however, *decisions* to allow drawn lines to "stand for" any line are not sufficient to establish that a particular property of the former is a property of any of the latter.

More appropriately one might say that the demonstration shows that any (x) qua line *must* have a certain property. The perceptual configuration in the diagram (the perceptually identified or constructed "line") is an aid to showing this, and letting this configuration "stand for" all lines is elliptical for saying, I plan in the demonstration to articulate the concept of "a line in general." The demonstration then offers no assurance that the figure in the diagram, or any other perceptual figure, is a "line." It is in this sense that the demonstration has no existential import. Yet, on two previously discussed grounds, Berkeley appears to reject this view.

---

[19] On this point see Wild, *op. cit.,* pp. 386-394. Wild evidently thinks Berkeley has solved the problem in the *Analyst,* although it is not clear how he arrives at this view.

The notion of explicating or articulating the concept of geometrical figures ("lines," "triangles," etc.) "in general," would suggest that the object of geometry is "general abstract ideas." This is particularly true since Berkeley understandably never doubts that geometry has an "ideational" content; that it is related to sense and imagination. In the second place (and related to the preceding remark) Berkeley believed that perceptual extension was the subject-matter of geometry. Geometry, then, is about space, but perceptual space, not the "absolute space" which he identifies with either the Cartesian's concept of extension simpliciter, or the Newtonian separately existing container of physical events and processes.

In closing this section we would note that we take issue with G. W. Ardley, who, in his essay, "Berkeley's Philosophy of Nature," makes the following statement:

the propositions of geometry [for Berkeley] are not about realities; they are general statements which have real significance only when referred to sensible particulars. The akribeia of the science pertains only to its phantom state; in the concrete, i.e. the real, the real situation is rough and ready; the akribeia serves as a frame of reference, or code of discipline, in human ordering of the concrete.[20]

Ardley evidently bases his position on passages like Section 14 of the *Essay on Vision,* where Berkeley does speak of the "lines and angles" used in geometrical optics to judge distance, as having no existence, being merely a "hypothesis" introduced into "optics." However, there is no unequivocal interpretation of this and comparable passages (*De Motu* 39, 61). The concept of "fictitious entity" in geometry may have a much more limited extension for Berkeley, merely referring for some practical purpose to a situation that does not exist. Section 61 of *De Motu* speaks for example of a "curve . . . considered as consisting of an infinite number of straight lines, though it in fact does not consist of them. "Fictitious," here means simply attributing a property to something, when it does not have the property. Similarly with the "lines and angles" in geometrical optics, the claim may merely be that there are no "lines" from the visually perceived object to the eye, and not the claim that the term "line" in "pure" geometry refers to some ideal unperceived object. Secondly, Ardley's thesis does not come to grips, in our judgment, with the continual assertion *later* in the *Essay,* and in the *Principles* as well, that perceptual extension is the object of geometry.

---

[20] G. W. Ardley, *op. cit.,* p. 22.

Our third objection to Ardley's position could be expressed in the following way. If Berkeley in fact distinguishes "pure" and "applied" geometry, we can ask what he would view to be the object (referents) of the primitive geometrical terms, e.g. "points," "lines," etc. The answer could not be, there are no referents for the primitive terms, for this would make pure geometry a science purely "verbal," a point of view never expressed by Berkeley. If we answer that the objects of "pure" geometry are ideal "lines," "points," etc. which although unperceived are said to perfectly embody the Euclidean axioms, we have the problem that such objects, viewed in some sense as constituting a content of consciousness and therefore "ideas," would be unacceptable to Berkeley, for like Locke's "general triangle," they would have the status of "abstract general ideas." The third possibility concerning the "objects" of the geometrical primitives, is that they are perceptual "lines," "points," etc. This possibility seems to us most consistent with the Berkeleyian texts, although if true, would imply that there is for Berkeley no distinction, of the sort Ardley suggests, between pure and applied geometry.

THE CRITIQUE OF ANALYSIS

A. The Concept of Infinity

Berkeley's concern with the concept of infinity in mathematics is apparent throughout his writings. His conception or view that perceptual extension is composed of aggregates of sensible minima is presented in conscious opposition to what he took to be the traditional view, that finite segments of extension were infinitely divisible. Our own previous discussion suggested that Berkeley's sensible arithmetization of perceptual extension can be approached from two points of view. (1) The doctrine that finite extension is infinitely divisible violates the *conjoined* principles that (a) it is intuitively obvious that there is a finite minimal perceptual expanse, and (b) there is no space (or extension) other than perceptual space (or extension). (2) The concept of finite magnitudes actually divided into an infinite number of parts raises those alleged paradoxes of the actual infinite, some of which are known as „Zeno's paradoxes."

The latter (Zenonian paradoxes) stem not from the claim that a finite segment can be potentially divided infinitely, which may, after all, take an infinite amount of time, but rather stem from the supposition that an infinite number of tasks can be accomplished in a finite time. Assuming the infinite divisibility of the segment, Achilles must traverse, in a finite time, an infinite number of non-overlappng segments before overtaking the tortoise.

Berkeley was aware of the distinction between the potential and the actual infinite, and in an interesting early essay, "On Infinites," he makes use of it.[21]

Berkeley begins the essay by quoting with approval a passage from Locke, where the distinction is made:

I guess we cause great confusion in our thoughts when we joyn infinity to any suppos'd idea of quantity the mind can be thought to have, and so discourse or reason about an infinite quantity, viz. an infinite space or an infinite duration. For our idea of infinity being as I think an endless growing idea, but the idea of any quantity the mind has being at that time terminated in that idea, to join infinity to it is to adjust a standing measure to a growing bulk; and therefore, I think 'tis not an insignificant subtilty if I say we are carefully to distinguish between the idea of infinity of space and the idea of space infinite.[22]

It is not clear again whether having an "idea" means to imagine or to find logically conceivable. The latter is the more plausible sense, for clearly the potential infinite, the possibility of a successive and infinite number or additions or divisions, is not an "idea" in the Berkeleyian sense. An inexhaustible sequence of operations is not an object of sense.

What is interesting is that Berkeley adduces the potential infinite as an argument against the conception of an actually infinitely small segment, or "infinitesimal":

Tis plain to me we ought to use no sign without an idea answering it and tis plain that we have no idea of a line infinitely small, nay tis evidently impossible there should be any such thing, for ever line, how minute so ever, is still divisible into parts less then itself; therefore there can be no such things as a line quavis data minor infinitely small.[23]

We will return to Berkeley's critique of the "infinitesimal" when we deal with the *Analyst*. Berkeley's argument here is surprising, however, for one would think that the concept of the potential infinite (the "idea" of the "infinity of space") is not compatible with the view that extensive magnitudes are composed of sensible minima. The existence of such minima would seemingly disallow an inexhaustible division of a finite segment. Therefore we would disagree with the following comment of A. A. Luce in his introduction to the essay:

[21] "On Infinites," ed. A. A. Luce in *Berkeley, Works,* Vol. 4.
[22] Locke, *Essay Concerning Human Understanding,* Book II, Ch. 17, Sec. 7, quoted by *Berkeley, Works,* Vol. 4, *op. cit.,* p. 235.
[23] *Ibid.,* p. 236.

For if space be not only infinitely divisible (subjectively) but infinitely divided (objectively), if, that is, it consists of infinitely small parts, then the new principle, esse est percipi, breaks down, and infinitely divisible space may well house infinitely divisible matter.[24]

Yet, as we have seen, the existence of sensible minima would entail a limit to the division of a finite spatial manifold.

When we discuss the *Analyst*, we will see that Berkeley's major objection to the concept of the infinitesimal is no longer based on the latter's incompatibility with the concept of the potentially infinite division of a finite segment, but rather on the alleged unintelligibility of the concept of a fixed finite magnitude said to be less then any assignable magnitude; a "zero increment," to use the apt phrase of John Wisdom.[25] In the early essay we are discussing, this issue of the intelligibility of the concept of the infinitesimal itself is somewhat obscured, since it is discussed in the context of the problem of infinitesimals (differentials) of various orders.

They [writers on the Calculus] represent upon paper infinitesimals of several orders, as if they had ideas in their minds corresponding to these words or signs, or as it it did not include a contradiction that there should be a line infinitely small, and yet another infinitely less then it.[26]

This criticism, though perhaps correctly pointing to ambiguities in the concept of differentials of various orders, tends to obscure the fundamental thrust of Berkeley's criticism which concerns the concept of the infinitesimal itself, whether statically conceived as a fixed quantity, or kinematically conceived as a "nascent" or "evanescent" velocity, or dynamically conceived as an "impetus."

We should note that from a modern point of view, there is no necessary relation between an actually infinite number of divisions of a finite segment, and the existence of infinitesimals. One can speak of the unit length being the sum of the following infinite series: $\frac{1}{2} + \frac{1}{4} + \frac{1}{8} + \frac{1}{16} + \ldots + \frac{1}{2^n}$; where no individual term in the series has the status of an infinitesimal.[27] With respect to analysis, the question of the existence of infinitesimals generally arises when we are asked intuitively

[24] See Luce's introduction to the *Essay, Works,* Vol. 4, p. 235.

[25] John Wisdom, "The *Analyst* Controversy: Berkeley as a Mathematician," *Hermathena,* 53, p. 112.

[26] "On Infinites," *op. cit.,* p. 235. On the controversy referred to see, Carl B. Boyer, *The History of the Calculus and Its Conceptual Development,* (Hafner Pub. Co., 1949, reprinted New York: Dover Publishing Co., 1959) pp. 213-214.

[27] See Grunbaum, *Modern Science and Zeno's Paradoxes, op. cit.,* pp. 43-44. "The length of a finite interval (a, b) is the non negative quantity b - a regardless of whether the interval is closed, open, or half open." From this point of view, we can say the length of the segment is "1" although the sequence of subintervals have no last element.

to comprehend how one figure, for example a circle (defined ostensively), can be *actually* equivalent to a qualitatively distinct figure, a polygon with an infinite number of sides; or how a given property, for example, the area under a finite segment of a curved line, can be conceived as an *actual* sum of an infinite number of rectangular areas, the width of each rectangle being less than any finite magnitude; or how an instantaneous rate of change can be conceived as an *actually* achieved limit of an infinite sequence of ratios, the elements of which (numerators and denominators) are assignable finite magnitudes. The concept of the infinitesimal conjoins two properties, both of which appear to be required to comprehend intuitively the kinds of changes mentioned above. The infinitesimal being a quantity *less* then any assignable magnitude allows us to posit an actually infinite sequence of elements, whose sum, for example, is not infinite. The infinitesimal as an actual *quantity* (greater than zero), saves us from the absurdity of having our *sums* dissolve into a mere sum of zeros, or the problem of having our *ratios* reduce to the unintelligible $^0/_0$.

Now the concept of the potential infinite is historically connected with an alternative method for solving problems of tangents and areas, dealt with by the calculus: the Archimedian method of "exhaustion." And in this early essay, Berkeley appears to recognize this latter method as an acceptable way of avoiding the paradoxes entailed in positing the existence of infinitesimals. For example, he says:

the supposition of quantities infinitely small is not essential to the great improvements of the Modern Analysis. For Mr. Leibniz acknowledges his Calculis differentialis might be demonstrated reductione ad absurdum after the manner of the ancients; and Sir Isaac Newton in a late treatise informs us his method of Fluxions can be made out a priori without the supposition of quantities infinitely small.[28]

Berkeley, in the *Analyst,* criticizes the method of "fluxions" for not avoiding the "paradoxes" of the infinitesimal. What is interesting to note here is the suggestion that the legitimate method of analysis, the Archimedian method of exhaustion, is connected with the legitimacy in the use of the concept of the potential infinite. And it is this connection which explains the following final passage in the early essay:

Now I am of the opinion that all disputes about infinities would cease, and the consideration of quantities infinitely small no longer perplex Mathematicians, would they but joyn Metaphysics to their Mathematics and conde-

---

[28] The late treatise Berkeley refers to is Newton's *Quadraturum Curvarum,* 1704.

scend to learn from Mr. Locke what distinction there is betwixt infinity and infinite.[29]

The method of exhaustion, however, involves both a certain assumption and a particular method of proof. The assumption is that the difference between two properties, for example, the area of a circle and an inscribed polygon with variable sides, can be made as small as one wishes; no claim being made that the polygon ever becomes a circle. The method of proof is the *reductio ad absurdum,* showing that one cannot hold without contradiction *both* the principle of the infinite (potential) diminishment of differences *and* the denial of the proposition to be demonstrated.[30]

The method of exhaustion, depending as it does on the principle of a potentially infinite diminishment of differences, would appear, however, to be incompatible with the concept of sensible minima. Such minima constitute a limit to even the potential division of a finite segment, and to even the potential diminishment of differences between, for example, the circumference of a circle and the perimeter of an inscribed polygon, the length of whose sides is allowed to vary. It is interesting to note in this regard that in the *Analyst,* Berkeley does not substitute the method of exhaustion as an intelligible alternative for the use of infinitesimals, but rather what is termed the "compensation of errors." We will have occasion to deal with this method below.[31]

We now turn to Berkeley's more detailed treatment of the calculus in the *Analyst.*

We have noted that the locution "(x) is inconceivable," does not, in Berkeley's writings, have a univocal sense. It often appears to mean, (x) is not an "idea," or object of sense. We have suggested that this is a derivative meaning, and that the locution has the primary sense that the concept of "(x)" is logically inconsistent. In any event we will focus on Berkeley's critique of the Calculus from the point of view of the latter's logical consistency and not on whether simply there are sensible referents for terms such as "momentaneous increments," "nascent and evanescent quantities," "fluxions and infinitesimals of all degrees," etc.[32] It is the logical critique that is recognized as having the historical role of a catalyst in the clarification of the fundamental concepts of analysis.[33]

---

[29] "On Infinites," p. 238.

[30] Carl B. Boyer, *History of Mathematics.*

[31] However, see *Analyst,* (Query 53, p. 101).

[32] *Analyst* (49).

[33] For example: Boyer, *History of the Calculus,* op. cit., Chapter 4. John Wisdom, "The Analyst Controversy," op. cit. John Wisdom, "The Compensation of Errors in

Berkeley begins the *Analyst* with an accurate, though condensed exposition of Newton's attempt to clarify the foundations of analysis, by appeal to kinematical ideas rather than to the concept of the infinitesimal as a static element. The critical concepts are "fluent," "fluxion," "moments," and "prime and ultimate ratios of evanescent increments (or nascent augments)." [34]

Newton begins with the conception of magnitudes ("fluents") as generated by moving points; the rate of generation is the "fluxion" or velocity, and the amount of magnitude generated in an infinitely small segment of time is the "moment." The appeal to the intuition of continuous motion is paramount; so that the "moment" is not to be viewed statically as an infinitesimal, but that which is in the process of becoming a finite magnitude. As Boyer points out, [35] "time" was always explicitly present for Newton as the independent variable, so that strictly speaking a "moment" was the product of a generating velocity and an infinitely small increment of time. In modern analysis, "time" in this explicit sense is left out, and we speak in a somewhat atemporal sense of the rate of change of any variable with respect to any other. The insight of Newton, developed into the algorithms of the Calculus, is that, given the "fluxion" or generating velocity (the modern derivative) one can determine the "fluent," and vice versa.

Berkeley correctly recognizes (*Analyst* 4) that the "fluxions" as velocities' ("celerities") cannot be viewed as the ratio or comparison of finite quantities, but, as he says, "only to the moments or nascent increments, whereof the proportion alone, and not the magnitude, is considered." The phrase "the proportion alone and not the magnitude" is significant, suggesting what becomes a conceptual problem for Berkeley, of conceiving finite assignable ratios whose components, however (numerator and denominator), cannot have any assignable magnitude. "Velocity", never loses for Berkeley the sense of being a comparison of finite assignable

the Method of Fluxions," *Hermathena*, No. 57, 1941, pp. 49-81.

Florian Cajori, "Discussion of Fluxions: From Berkeley to Woodhouse," *American Mathematical Monthly*, XXIV (1917), pp. 145-154. Florian Cajori, "The History of Zeno's Arguments on Motion," VI, "Newton, Berkeley, Jurin, Robbins and Others," *The American Mathematical Monthly*, Vol. XXII, May, 1915, No. 5, pp. 143-149.

[34] Newton's works in order were: *De Analysi per Aequationes Numero Terminorum Infinitas* (written 1669, published 1711). *Methodus Fluxionum et Serierum Infinitarum,* (written 1671, published 1736). *De Quadratura Curvarum* (written 1676, published 1704). First published results of Newton's discoveries in the Calculus appeared in the *Principia* 1687. Lemmas in the first book express the viewpoint of *De Quadratura Curvarum*. The general method for differentiation appears in the Second Book of the *Principia*. See Boyer, *The History of the Calculus, op. cit.*, pp. 189-202.

[35] *Ibid.*, p. 196.

magnitudes, a view that makes it difficult for him to appreciate Newton's discussion of "ultimate ratios" in Book 1 of the *Principia*. We will shortly return to a more detailed discussion of this issue.

Very early in the *Analyst*, Berkeley grasps the fact that the weakness in Newton's kinematic revision of the Calculus concerns the concept of a "moment."

Now, as our sense, is strained and puzzled with the perception objects extremely minute, even so the imagination, which faculty derives from sense, is very much strained and puzzled to frame clear ideas of the least particles of time, or the least increments generated therein: and much more so to comprehend the moments, or those increments of the flowing quantities. (*Analyst* 4)

Although the passage raises the problem in terms of the possibility of sensory awareness, it becomes clear that this is a derivative problem; more fundamental is whether the concept of "moment" is logically more intelligible than that of "infinitesimal." We still apparently have the problem of comprehending spatial or temporal increments greater than zero, but less then any assignable finite magnitude ("in status nascenti"). The appeal to a principle of continuity, through the intuition of motion, allows us to speak of an incipient motion, or a motion just generated, and thus somehow avoids the conception of finite magnitudes as aggregates of fixed infinitesimals. If we think of Achilles and the tortoise the difference in view is whether we think of Achillis somehow successively occupying every position in an infinite set, or whether he is to be said to be "passing" through each position. From the latter point of view, we think of the motion as "real," and the occupancy of a given point as a conceptual limit. The concept of a "conatus," as used, for example, by Leibniz, suggests that the continuity of motion requires that we attribute movement (velocity) to the body at every point, although clearly this "instantaneous" velocity is not (though it may be proportional to) a comparison of finite magnitudes.[36]

Berkeley's rigorous "either or," however, finds the appeal to the continuity of motion insufficient.

If by a momentum you mean more than the very initial limit, it must be either a finite quantity or an infinitesimal. But all finite quantities are expressly excluded from the notion of a momentum. Therefore the momentum must be an infinitesimal. And indeed, though much artifice hath been employed to escape or avoid the admission of quantities infinitely small, yet

[36] Leibniz, "Specimen Dynamicum," *op. cit.*, Boyer, *History of the Calculus, op. cit.*, pp. 215-216.

it seems ineffectual. For ought I see, you can admit no quantity as a medium between a finite quantity and nothing, without admitting infinitesimals. (*Analyst* 11)

In the concept of the "moment" the paradoxical "infinitesimal" appears in the notion of an "infinitesimal" portion of time.

An increment generated in a finite particle of time is itself a finite particle; and cannot therefore be a momentum. You must therefore take an infinitesimal part of time wherein to generage your momentum. (*Analyst* 11)

Berkeley's critical comments in this part of the *Analyst* are directed at Newton's account of the differentiation of a product, found in Book 2 of the *Principia;* [37] an account Newton himself considered to express the "foundation" of his method as expounded in the *Methodus Fluxionum*.[38] Newton considers the quantities he deals with ("products, quotients, roots, rectangles, squares, cubes... and the like") as

variable and indetermined, and increasing or decreasing as it were, by a continual motion or flux; and I understand their momentary increments or decrements by the name of moments. ... But take care not to look upon finite particles as such. Finite particles are not moments, but the very quantities generated by the moments. We are to conceive them as the just nascent principles of finite magnitudes.[39]

Berkeley, in our judgment, correctly notes that Newton's method for the differentiation of a product is not made clear in the *Principia*. Using modern terminology, if we consider the sides of a rectangle to be generated by the functions f(t) and g(t), (Newton's A and B) and the derivatives of each function to be f'(t) and g'(t) (Newton's a and b), then the derivative of the product according to Newton's Lemma is f'(t)g(t) + g'(t)g(t) (in Newton's terminology aB + bA). Newton, however, gains his result by considering the product when the sides have been "augmented" by one half their "moments," (A + ½a) (B + ½b) and substracting from this the product when the sides "want" one half their moments (A — ½a) (B — ½b), giving the desired result, aB + bA. Berkeley notes that the product AB should be subtracted from the product (A + a) (B + b), raising the issue of on what grounds the quantity ab (product of the "moments") should be ignored. More generally Newton offers no justification for taking a decrement of ½ a moment and subtracting it from

[37] Newton, *Principia, op. cit.,* BK II, Sec. II, Proposition 7, Theorem 5, Lemma 2, p. 168.
[38] Ibid., Scholium to Lemma 2, p. 170. See Boyer, *History of the Calculus, op. cit.,* pp. 198-199.
[39] Newton, *Principia, op. cit.,* p. 168.

an increment of $\frac{1}{2}$ a moment, except that $\frac{1}{2}a - (-\frac{1}{2}a) = a$. However $\frac{1}{4}a - (-\frac{3}{4}a)$ also equals a, but if this particular increment and decrement were chosen, the lemma would not have been demonstrated.

it is not easy to conceive why we should take quantities less then A and B in order to obtain the increment of AB, of which proceeding it must be owned the final cause or motive is very obvious; but it is not so obvious or easy to explain a just and legitimate reason for it, or show it to be geometrical. (*Analyst* 11)

Berkeley then turns his attention to Newton's determination of the fluxion of X^n in *De Quadratura Curvarum*, where the latter attempted, in the words of Boyer, "to remove all traces of the infinitely small." [40] Briefly the method is as follows: representing the variable "X" as augmented by the increment "O;" to expand (by the binomial expansion) $(X + O)^n$; substracting X^n from this result: and forming the ratio of the "increments $(X + O)^n - X^n$ and $(X + O) - X$, divide by "O" resulting in the ratio 1 *to*

$$\frac{nx^{n-1}0}{1} + \frac{(n-1)(n)x^{n-2}0^2}{2.1} + \ldots + 0^n.$$

As "O" approaches zero, what Newton terms the "ultimate ratio of the evanescent increments" of the "fluxion" of X^n becomes 1 *to* nX^{n-1}.[41] (*Analyst* 13)

We will shortly consider more fully Newton's concept of an "ultimate ratio." It seems clear that Berkeley interprets it as an actual comparison of magnitudes when the increment "O" vanishes; and this he legitimately finds to be contradictory.

Hitherto I have supposed that X flows, that X hath a real increment, that "O" is something . . . From that supposition it is that I get at the increment of X^n, that I am able to compare it with the increment of X, and that I find the proportion between the two increments. I now beg leave to make a new supposition contrary to the first, i.e., I will suppose that there is no increment of X, or that "O" is nothing; which second supposition destroys my first, and is inconsistent with it; and therefore with everything that supposeth it. All of which seems a most inconsistent way of arguing, and such as would not be allowed of in Divinity. (*Analyst* 14)

Berkeley also correctly perceives that, with respect to the logic of the demonstration, there is no difference between ignoring a static infinitesimal and the kinematically conceived "evanescent increment."

[40] Boyer, *History of the Calculus, op. cit.*, p. 195.
[41] *Ibid.*

... and indeed it requires a marvellous sharpness of discernment to be able to distinguish between evanscent increments and infinitesimal differences. It may perhaps be said that the quantity being infinitely diminished becomes nothing, and so nothing is rejected. But according to the received principles it is evident that no geometrical quantity can by any division or subdivision whatsoever be exhausted, or reduced to nothing. (*Analyst* 17)

Berkeley rightly observes that Newton, as demonstrated by the various formulations of his method, appeared to be unclear about the meaning of his fundamental concepts. Before turning to a more detailed consideration of one particular concept, that of an "ultimate ratio of evanescent increments," we will consider Berkeley's claim that the results of differentiation can be achieved through the method of "compensation of errors," a method involving only finite considerations.

We will follow Wisdom's classification of two types of "compensation": (A) an analytic compensation for an analytic error; and (B) the balancing of an analytic error by a compensating geometrical error.[42] Our claim will be that, as presented by Berkeley, the "compensation of errors" is not a legitimate mathematical method, and could not substitute as a valid procedure for the apparently invalid method of making use of and then ignoring certain increments in differentiations.

In our remarks, we will first refer to the following diagram from the *Analyst*. (Section 21)

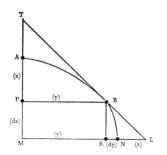

Abscissa   AP = x
Ordinate PB = y
PM = dx (difference in the abscissas)
RN = dy (difference in the ordinates)
NL = z

If we ignore (z) and consider the "differential triangle" BRN similar to the actual triangle TBP, we get the proportion $dy/dx = y/PT = y\,dx/dy$. However, for any (z) greater than zero, (considering the true triangle BRL) the proportion should read $PT = ydx/[dy + z]$ Berkeley's statement (*Analyst* 21);

There was therefore an error of defect in making the divisor: which error was equal to z,

[42] Wisdom, "The Compensation of Errors in the Method of Fluxions," *op. cit.*

is misleading. If by "error of defect" is meant the difference in the two results, this is not z, but (ydx) (z) / (dy) (dy + z).

Berkeley then claims that if we represent the curve AN by the equation $y^2 = px$, the algorithm of the calculus is said to give us, $dy = pdx/2y$. His contention is that this result ignores the quantity $dy^2/2y$. This is because if we take $(y + dy)^2 = p(x + dx)$ and solve for dy, we get dy = $[pdx/2y]$ — $dy^2/2y$. Since $dy^2/2y$ can be shown to be equal to z; Berkeley claims:

the two errors being equal and contrary destroy each other; the first error of defect corrected by the second error of excess. (*Analyst* 21)

Yet, it is difficult to see where the "compensation" lies. If we solve in the first part we have dy = $[ydx/PT]$ — z, showing that the desired result, dy = ydx/PT contains an excess of z units. In both cases, then the *desired result contains an excess of z.*

In the second case, where there is an alleged compensation of an analytic with a geometrical error, our remarks will refer to the following diagram. (Section 28 — *Analyst*)

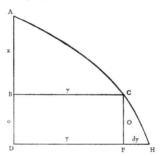

$y = x^n$
"o" = an increment in x.
"dy" = an increment in y.

Berkeley's argument is as follows:

Let the Area ABC be equal to x^n. By the method of fluxions, the "ordinate" (y) is said to be equal to nx^{n-1}.

The reason for this result "plainly appears," if we divide the incremental area BCDH into two parts, BDFC + CFH. We note that since area ABC = x^n then the remainder in the expansion of

$(x+o)^n$ or $nox^{n-1} + \dfrac{n(n-1)o^2x^{n-2}}{2} + \ldots + 0^n$ must equal BDFC + + CFH.

We then retain only the first members of each side of this equation, getting nox^{n-1} = BDFC. If we then divide by "o" = BD, we get the desired result that nx^{n-1} = BC (the ordinate or y) with no illegitimate suppression of allegedly infinitely small quantities.

However, the validity of the above procedure requires that the differential triangle CFH be set equal to $\dfrac{n(n-1)o^2 x^{n-2} + \ldots + 0^n}{2}$,

"agreeable to the axiom that if from equals you subduct equals, the remainders will be equal." (*Analyst* 28) Yet Berkeley offers no independent justification for this claim of equality, beyond the fact that if the claim is made, the desired result is obtained. For example, in section 28, he comments:

If therefore the conclusion be true, it is absolutely necessary that the finite space CFH be equal to the remainder of the increment expressed by $(n^2 - n/2)$ (o^2) (x^{n-2}) & C.

Again, however, there is no suggestion that the "equality" can be independently demonstrated.

We are faced, then, with the question of whether Berkeley believed the "compensation of errors" to be a mathematically correct alternative to the method of "fluxions." The next section in the *Analyst* (Section 29), although not unambiguous, suggests that he did believe this, and suggests, we well, the source of Berkeley's error. (Refer to the diagram on p. 180)

Therefore, be the power what you please, there will arise on one side an algebraic expression, on the other a geometrical quantity, each of which naturally divides itself into three members. The algebraical or fluxionary expression, into one which includes neither the expression of the increment of the absciss nor of any power thereof; another which includes the expression of the increment itself; and a third including the expression of the powers of the increment. The geometrical quantity also or whole increased area consists of three parts or members, the first of which is the given area; the second a rectangle under the ordinate and the increment of the absciss; and the third a curvilinear space. And, comprising the homologous or correspondent members on both sides, we find that as the first member of the fluxionary expression is the expression of the given area, so the second member of the expression will express the rectangle or second member of the geometrical quantity; and the third containing the power of the increment, will express the curvilinear space, or third member of the geometrical quantity.

Symbolically: $\overset{1}{\overbrace{ABC}} + \overset{2}{\overbrace{BCDF}} - \overset{3}{\overbrace{CFH}} = \overset{1}{\widetilde{x^n}} + \dfrac{\overset{2}{\overbrace{nx^{n-1}0}} + \ldots + \overset{3}{\overbrace{0^n}}}{1}$.

However, again, Berkeley offers no demonstration for his claim that the "second member" of the fluxionary expression

$$\frac{(nx^{n-1}o)}{1}$$

should be "homologous" with the rectangular area BDFC (yo). We can speculate that Berkeley was misled by a belief that if the quantities, $a + b = c + d$, then $a = c$, and $b = d$. The method Berkeley presents, then assumes rather than demonstrates the very point at issue: that the rectangular area yo is equal to the expression $\dfrac{nx^{n-1}o}{1}$.

We now return to Berkeley's critique of Newton's attempt, in his method of fluxions, to eliminate considerations of infinitely small quantities, or infinitesimals.[43]

Berkeley's fundamental argument is that the logical problems with the concept of the infinitesimal are not solved by intuitive appeal to the continuity of motion. Although Newton's "moment" is not statically conceived as a fixed but infinitely small quantity, it is, according to Berkeley, either something or nothing; in the former case, results presaged on having "moments" actually disappear (take on the value zero) are illegitimate, in the latter case, the quantity represented by the "moment," the product of a generating velocity and an infinitesimal increment of time would of necessity be zero. The problem of the infinitesimal emerges, as Berkeley recognized, in the very conception of an infinitesimally small increment of time. The intelligibility of the concept of a "moment" as a generated magnitude, less then any assignable magnitude, supposes that the concept of a "nascent" or just beginning temporal interval (less then any assignable magnitude) is intelligible.

It should be noted, in this regard, that in the *Analyst* Berkeley's rejection of temporal and spatial infinitesimals in not, as it is in the *Principles* (sec. 123-125), based on his doctrine of sensible minima; although the latter conception is incompatible with claims for the existence of infinitesimals (as it is incompatible with the claim for any finite quantities less then the minimum). The reiterated point of view of the *Analyst* is that the concept of a quantity said both to be greater than zero and less then any finite assignable magnitude is unintelligible. Berkeley's objections to the concept of "velocity at an instant" are primarily of a logical nature. "We have, he says,

no notion whereby to conceive and measure various degrees of velocity beside space and time; or when the times are given, beside space alone. We

[43] *Analyst* (29) "Why then are fluxions introduced? Is it not the show or rather to palliate the use of quantities infinitely small?"

have even no motion of velocity presinded from time and space. When therefore a point is supposed to move in given times, we have no notion of greater or lesser velocities, or of proportions between velocities, but only of longer and shorter lines, and of proportions between such lines generated in equal parts of time. (*Analyst* 30)

"Prescinding" the notion of "velocity" from "space and time" is not fundamentally illegitimate because an instantaneous velocity cannot be an object of sense or measurement, but because the very concept of a velocity entails the comparison of finite spatial and temporal magnitudes. A velocity is a ratio, and a ratio can only be conceived as a comparison of two distinct and assignable magnitudes.

For, to consider the proportion or ratio of things implies that such things have magnitude. (*Analyst* 31)

From this point of view, we would disagree with Boyer, who remarks:

Berkeley, the extreme idealist, wished to exclude from mathematics the "inconceivable" idea of instantaneous velocity. This is in keeping with Berkeley's early sensationalism, which led him to think of geometry as an applied science dealing with finite magnitudes which are composed of indivisible "minima sensibilia." [44]

In our judgment, however, the concept of sensible minima plays, in the *Analyst*, little or no role in Berkeley's rejection of both infinitesimals and instantaneous rates of change. It is in fact the unintelligibility of the concept of the infinitesimal, conjoined with the view that a velocity must be a ratio or comparison of real magnitudes, that engenders Berkeley's rejection of the legitimacy of an instantaneous velocity.

The issue in historical context concerned how one was to understand Newton's "ultimate ratios," and Berkeley was certainly not unique in finding a lack of clarity in the concept.[45] Berkeley evidently entertains the possibility of conceiving of these ratios (the modern derivative) as "limits," but his conception of "limit" appears to be that of a geometrical boundary.

A point may be the limit of a line. A line may be the limit of a surface: a moment may terminate time. But how can we conceive velocity with the help of such limits. It [velocity] necessarily implies both time and space, and cannot be conceived without them. (*Analyst* 31)

---

[44] Boyer, *History of the Calculus, op. cit.*, p. 227.
[45] *Ibid.*, pp. 228-230, Discussion of the Views of James Jurin and Benjamin Robbins.

In Berkeley's reply to Walton [46] we find this same geometrical conception of a "limit":

If this able vindicator should say that quantities infinitely diminished are nothing at all, and consequently that, according to him, the first and last ratios are proportions between nothings, let him be desired to make sense of this, or explain what he means by proportions between nothings. If he should say, the ultimate proportions are the ratios of mere limits, then let him be asked how the limits of lines can be proportioned or divided.

Since the limit of a line, or a point, has the magnitude zero, Berkeley correctly finds unintelligible the notion of an "ultimate ratio" being an actual comparison of limits.

It is this overriding geometrical view of magnitude which perhaps prevents Berkeley from entertaining the conception of an "ultimate ratio," not as a ratio of geometrical limits, but as a numerical limit to which a sequence of actual ratios (comparison of magnitudes) converges. And although Newton's discussion of ultimate ratios is not unambiguous, the latter conception appears clearly to be suggested. We refer here to certain lemmas and scholia in Book 1, Section 1 of the *Principia* (particularly the interpretive scholium after lemma 11).[47] For example, in lemma 1, Newton states:

Quantities, and the ratios of quantities, which in any finite time converge continually to equality, and before the end of that time approach nearer to each other then by any given difference, becomes ultimately equal.[48]

This lemma is then made use of in demonstrating certain relationships between rectilinear and curvilinear figures.[49] The scholium at the end of lemma 11, however, makes clear that expressions such as "ultimate equality," "ultimate form" (of the "evanescent triangles"), and "ultimate ratio," are a *façon de parler*, and not to be taken literally. We quote a substantial portion of the scholium to give a sense of Newton's conception of "ultimate ratios" as limits:

These lemmas are premissed to avoid the tediousness of deducing involved demonstrations ad absurdum, according to the method of the ancient geometers. For demonstrations are shorter by the method of indivisibles; but because the hypothesis of indivisibles seems somewhat harsh, and therefore that method is reckoned less geometrical, I chose rather to reduce the demonstrations of the following propositions to the first and last sums and ratios

    [46] Berkeley, "Concerning Mr. Walton's Vindication of Sir Isaac Newton's Principles of Fluxions," appendix to the "Defense of Free Thinking in Mathematics."
    [47] Newton, *Principia, op. cit.*, pp. 25-32.
    [48] *Ibid.*, Lemma 1, p. 25.
    [49] *Ibid.*, Lemma 8, pp. 27-28 as an example of Newton's method.

of nascent and evanescent quantities, that is to the limits of those sums and ratios, and so the premiss as short as I could the demonstration of those limits ... Therefore if hereafter I should happen to consider quantities as made up of particles ... I would not be understood to mean indivisibles, but evanescent divisible quantities; not the sums and ratios of determinate parts, but always the limit of sums and ratios.

In what sense the concept of limit is understood is expressed at the end of the scholium.

It may also be objected, that if ultimate ratios of evanescent quantities are given, their ultimate magnitudes will also be given: and so all quantities will consist of indivisibles, which is contrary to what Euclid has demonstrated concerning incommensurables, in the tenth book of his elements. But this objection is founded on a false supposition. For those ultimate ratios with which quantities vanish are not truly the ratio of ultimate quantities, but limits towards which the ratios of quantities decreasing without limit do always converge; and to which they approach nearer than by any given difference, but never to beyond, nor in effect attain to, till the quantities are diminished in infinitum.

Some important points should be noted about the sense of the quoted passages. (1) The "ultimate ratio" is not an actual comparison of magnitudes, finite or infinitesimal, but rather the numerical limit to which an actual sequence of ratios converges. In this sense, although Newton is not explicit on this point, the "ultimate ratio" is not a member of the set of actual ratios (actual comparison of magnitudes) for which it is the numerical limit. (2) The claim that the difference between the actual ratios and the limit can be made as small as one wishes, entails that the set of actual ratios is infinite. (1) and (2) together raise the question, debated by both critics and defenders of the Newtonian conceptions, whether a variable can be said to "reach" its limit.[50] Given a temporal understanding of the process of "reaching," the question is, in fact, an expression of the Zenonian kinematic "paradox" of how an infinite number of successive and non-overlapping states could be achieved in a finite time.

Newton, as far as we know, does not discuss the Zenonian paradoxes; but his answer to the question of whether variables (for example, velocities) can "achieve" their limits in a finite time seems clearly affirmative. In the Scholium, previously quoted, he remarks:

Perhaps it may be objects that there is no ultimate proportion of evanescent quantities; because the proportion, before the quantities have vanished, is not the ultimate, and when they are vanished is none. But by the same

[50] Boyer, *History of the Calculus*, pp. 229-232.

argument it may be alleged that a body arriving at a certain place, and there stopping, has no ultimate velocity; because the velocity, before the body comes to the place, is not its ultimate velocity; when it has arrived there is none. But the answer is easy; for by the ultimate velocity is meant that with which the body is moved, neither before it arrives at its last place and the motion ceases, nor after, but at the very instant it arrives; that is, that velocity with which the body arrives a its last place, and with which the motion ceases. And in like manner, by the ultimate ratio of evanescent quantities is to be understood the ratio of the quantities not before they vanish, nor afterwards, but with which they vanish. In like manner the first ratio of nascent quantities is that with which they begin to be.[51]

Briefly, if a body is *said* to be in motion (as opposed to rest) at time (t) it has a velocity at time (t). This is its *instantaneous velocity*. The magni-

[51] Newton, *Principia, op. cit.,* p. 31. See also *Quadratura Curvarum,* Whiteside ed., *Mathematical Works,* Vol. 1, p. 141. We refer to the diagram below.

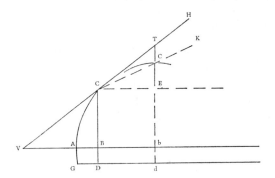

Newton comments: "Tis the same thing if the Fluxions be taken in the ultimate Ratio of the Evanescent parts." The "evanescent triangle" (or "mixtilinear" triangle) cEc in its "ultimate form" becomes similar to trainge CET, when, Newton suggests "C and c exactly coincide." "For," he says, "Errors, tho never so small, are not to be neglected in Mathematics." Berkeley quotes the latter passage with approval (*Analyst* 9) and turns it against Newton's own demonstration in Book 2 of the *Principia,* of gaining the fluxion of the product of two variables. Berkeley correctly notes concerning the above diagram (*Analyst* 34) that if "c" and "C" do actually coincide, we no longer have the "evanescent triangle" CcE, and the two linear triangles CTE and CcE. This is compatible, however, with saying that the "ultimate ratio" conceived as a limit of the sequence of ratios cE/CE is equal to the ratio of Tb/Vb. Again however this ultimate ratio could not be conceived as an actual comparison of finite magnitudes (though it may be set proportional to such an "actual" ratio) just as an instantaneous velocity cannot be conceived as an actual ratio of some distance interval to a temporal interval. Berkeley, then, misunderstands Newton when he (Berkeley) says: with reference to the diagram. "A point therefore is considered as a triangle, or a triangle is supposed to be formed in a point. Which to conceive seems quite impossible." (*Analyst* 34) Newton might allow that when "c" actually (in a "finite" time) coincides with "C," there are no triangles CEc, CEc and CTE, and the "ultimate ratio," (limit) of cE/CE is equal to the ratio of Tb/Vb. If Newton means to say that when "c" coincides with "C" that we have actually similar *triangles* CcE and CET, Berkeley's criticism is correct, that this is meaningless.

tude of this velocity is the numerical limit of an infinite of "average" velocities or actual ratios of finite distances to finite times. This crucial point of the scholium, that certain variables "achieve" (are defined for) their limits, although these limits are not actual ratios (*comparison* of actual magnitudes), is somewhat obscured by one of Newton's own examples. Towards the end of the scholium he says:

> This thing will appear more evident in quantities infinitely great. It two quantities, whose distance is given, be augmented in infinitum, the ultimate ratio of these quantities will be given, namely, the ratio of equality; [52] but it does not follow from thence, that the ultimate or greatest quantities themselves, whose ratio that is, will be given. Therefore if in what follows, for the sake of being more easily understood, I should happen to mention quantities as least, or evanescent, or ultimate, you are not to suppose that quantities of any determinate magnitude are meant, but such as are conceived to be always diminished without end.[53]

In this example there is no claim that in any finite time the magnitudes become equal (the process of augmentation could go on indefinitely), and thus the example is not relevant to the program established by lemma 1, which is to speak of "quantities and the ratios of quantities, which in any *finite time* converge continually to equality." The *single* point Newton is making with the example is that the "quantities" said "ultimately" to be equal do not have assignable magnitudes. A comparable example could be used concerning the infinite diminishment of two variable quantities $3x$ and x, which can be said to approach equality (become "ultimately" equal) as x approaches zero, (in infinitely diminished) although they maintain a constant ratio (Newton's example has a constant difference). Both examples, however, can be said to appeal to the "potential infinite," that is to a temporally infinite process of augmentation or diminishment. They more clearly refer to the method of exhaustion than the example of an instantaneous velocity, concerning which, Newton's claim is that a body can achieve such a velocity in a finite time.

With Newton's arguments of Book 1 in the *Principia* in mind, Berkeley responds to James Jurin,[54] who had accused him of not paying attention to them. Berkeley remarks, "I had long since consulted and considered it," (*Defense* 32) and goes on to remark:

> For, however, that way of reasoning may do in the method of exhaustions, where quantities less then assignable are regarded as nothing; yet for a

---

[52] For example, as x approaches infinity (gets as large as one wishes) the expression $(3x + 2)/3x$ approaches 1.

[53] Newton, *Principia, op. cit.*, p. 32.

[54] Berkeley, "A Defense of Free Thinking in Mathematics," *op. cit.*

fluxionist writing about momentums to argue that quantities must be equal because they have no assignable difference seems the most injudicious step that could be taken. (*Defense* 32)

And, towards the end of the same section, with reference to the "first section of the first book of the *Principia*," Berkeley writes:

And, indeed who sees not that a demonstration ad absurdum more veterum, proceeding on a supposition that every difference must be some given quantity, cannot be admitted in, or consistent with, a method wherein quantities less then any given, are supposed really to exist, and be capable of division.

Berkeley's remarks are specifically directed against Newton's demonstration of the fluxion of a product in Book 2 of the *Principia,* where, as we have commented, the grounds for ignoring the product of the "moments" (ab) are not made clear. On the other hand, Berkeley appears to miss a fundamental point in the discussion in Book 1 of the *Principia:* that although the term "ultimate ratio" does not refer to a comparison of actual magnitudes, the numerical value of such "ratios" or limits *are* considered to be defined for some variables, for example, the velocity of a moving body. If we assume that a certain sequence of average velocities converges to a numerical limit (K) in, say, two seconds, we must either say that the velocity *at* t $=$ 2 is (K) or is undefined for that t. The scholium previously quoted demonstrates that for Newton, as long as the body is said to be moving at t $=$ 2, it has a velocity at t $=$ 2, and this is its instantaneous velocity. It is, then, somewhat strange that Berkeley, who finds the concept of an "instantaneous velocity" to be unintelligible, offers no objection to Newton's discussion of "ultimate ratios," in Book 1 of the *Principia.*

Berkeley would contend that the velocity is "undefined" at t $=$ 2, as it is undefined at any instant. However his grounds, at least in the *Analyst* and related writings are not that instantaneous velocities are incompatible with the existence of spatio-temporal sensible minima, but rather that it is part of the meaning of the concept of "velocity" that it involves a comparison of finite spatial and temporal magnitudes. Moreover Berkeley appears to interpret an "ultimate ratio" (or at least for the sake of argument accept the interpretation of Newton's defenders like Jurin) as an actual ratio when delta S and delta T (average distance and time) have become zero, giving the unintelligible $^0/_0$.[55]

---

[55] See Berkeley's reply to Walton; (Appendix to the *Defense*) "If this able vindicator should say that quantities infinitely diminished are nothing at all, and consequently that according to him, the first and last ratios are proportions between nothings; let him decide to make sense of this, or explain what he means by proportions between nothings."

It is interesting to speculate whether Berkeley would have been willing to attribute instantaneous velocity to a body, if there had been greater clarity concerning Newton's ultimate ratios – that they were not themselves members of the infinite sequences of ratios (average velocities) for which they were the numerical limit. The answer would clearly be no, if we meant that bodies in some *real* sense could have such velocities. The "denseness" of perceptual space and time is rules out by the doctrine of sensible minima; and immaterialism ruled out that there can be an ontological distinction between perceptual and "physical" space and time. And in rejecting the idea that a finite spatio-temporal interval (for example a body traveling four feet in three seconds) is dense (that we can construct in the interval an infinite number of average velocities), it makes no sense to speak of an "instantanteous velocity" as a limit for the sequences in such "dense" intervals. And since there is no direct measure of an instantaneous velocity, our ontology could extend no further than our sensible practice of measurement, which is limited, in the case of the direct measure of velocity, to comparing finite spatial and temporal intervals.

Would Berkeley, on the other hand, be willing to accept an instantaneous velocity as a "mathematical fiction" (no phenomenal reference) useful in the formulation of the mathematical laws of physics? There are suggestions throughout Berkeley's works that he allows something like this. In *De Motu* (Section 18) he speaks of the usefulness of the law of the parallelogram of forces: "they serve the purpose of mechanical science and reckoning"; but they do not necessarily refer to anything in reality (do not "set forth the nature of things"). However, the context of this passage, particularly the previous section, strongly suggests that the "mathematical fiction" here is not the alleged independent action of forces, but merely the reference to something called "force." And Berkeley has never made it clear in what sense he means the reference to "forces" to be useful in mechanics. In *De Motu* (61), a passage previously referred to, Berkeley acknowledges that a circular motion can be considered as "arising from an infinite number of rectilinear directions," a conception, he says, "useful in mechanics." Again Berkeley does not go into detail as to how it is useful: that it appeals to the composition of forces, in this case inertial motion along the tangent and an acceleration towards the center. In fact it was this conceptualization of circular motion which was crucial in the development of the law of gravitation.

Comparable to the above examples, one could give instances where the assumption of an instantaneous velocity is useful in mechanics: in the concept of an "initial velocity," which enters into laws which predict the horizontal and vertical distance traveled by projectiles; in the concept of an "impact velocity," which enters into laws which predict the penetration of a projectile into a particular medium; etc. Making use of such concepts allows us to predict consequences that can be ascertained or checked by direct measurement. And one could argue that if one could not make use of such concepts, physics, and hence our powers of prediction, would be severely restricted.

It should be noted that if Berkeley had considered the use of "instantaneous velocity" a useful fiction, it would have to be a derivate fiction. Since instantaneous rates of change are computed as the numerical limits for infinite sequences, the more fundamental "fiction" is the assumption of "denseness" for finite spatial and temporal intervals. Can we conclude that Berkeley would hold that although perceptual, and hence real extension and duration, are composed of minima, so that finite spatial and temporal intervals cannot be infinitely subdivided, yet allow such subdivision as a "fiction" that is useful in the formulation of the laws of physics?

It seems correct that a general case could be made for such a "fictionalist" view. We cannot in any direct way demonstrate the "denseness" of physical space and time.[56] Moreover, any experimentally arrived at list or sequence of relative velocities cannot be established (merely by inspection) as necessarily converging towards some numerical limit. "Instantaneous velocities," like "mass points," might be considered as idealizations which allow us to simplify our expression of physical laws, but refer to nothing that can be said to be experimentally ascertainable.[57]

On the other hand there are numerous passages in Berkeley's writings that suggest the "denseness" of the space and time in which physical events take place is not allowable even as a useful "fiction." The passages in the *Principles* (Section 123-124) which claim that the doctrine of "infinite divisibility" of finite extension is the source of "those amusing geometrical paradoxes which have such a direct repugnancy to the plain common sense of mankind." We have suggested that Berkeley may have had in mind the Zenonian paradoxes (among others) concerning the actual infinite. For example, Newton's clear suggestion in Book 1 of the *Principia* that variables can "reach" their limits in a finite time, raises the spectre of the Zenonian paradoxes of motion like the "Achilles." And

---

[56] See Grunbaum, *Geometry and Chonometry in Philosophical Perspective, op. cit.*
[57] On the question of whether "velocity at an instant" can be legitimately predicated of actual bodies, see the discussion by Ernest Nagel, *The Structure of Science, op. cit.*, Ch. 6, "The Cognitive Status of Theories."

although Newton does not discuss the problem, the issue of whether variables can reach their limits becomes a central focus in the controversy engendered by Berkeley's *Analyst*.[58] And, although Berkeley does not specifically discuss the problem, it is not clear how such "paradoxes" would be obviated by allowing as a mathematical fiction the denseness of finite segments of space and time. Berkeley makes clear that the doctrine of "sensible minima" freed *"geometry"* from the paradoxes (our emphasis):

Hence, if we can make it appear that no finite extension contains innumerable parts, or is infinitely divisible, it follows that we shall at once clear the science of geometry from a great number of difficulties and contradictions which have ever been esteemed a reproach to human reason ... (*Principles* 123)

This point is echoed in the "queries" Berkeley attached to the end of the *Analyst*. For example, query 1:

Whether the object of geometry be not the proportions of assignable extensions? And whether there be any need of considering quantities either infinitely great or infinitely small?

and query 5:

Whether it doth not suffice, that every assignable number of parts may be contained in some assignable magnitude? And whether it be not unnecessary, as well as absurd, to suppose that finite extension is infinitely divisible.

The passages are not unequivocal in meaning, but we are taking them to suggest that for Berkeley, geometry (in which he includes analysis) has no need of the concept of "denseness" for finite spatial and temporal intervals. Berkeley, as we have pointed out, often suggests that unlike algebra (and perhaps arithmetic *considered* merely as a formalism) geometry in its very formulations is limited by the nature of its subject-matter, which is perceptual extension.

Whether because, in stating a general case of pure algebra, we are at full liberty to make a character denote either a positive or negative quantity, or nothing at all, we may therefore in a geometrical case, limited by hypotheses and reasonings from particular properties and relations of figures, claim the same license. (*Analyst* query 27)

The above passage strongly suggests that the mathematical fiction of

[58] See Boyer, *Concepts of the Calculus, op. cit.*, p. 231. Also see John Wisdom, "The Analyst Controversy – Berkeley's Influence on the Development of Mathematics," *Hermathena*, No. 29, 1939, pp. 3-29.

"denseness" would not be allowed by Berkeley. On the other hand, it must be remembered that the central focus of the *Analyst* is the internal incoherence of such concepts in the "modern analysis" as "infinitesimals," "moments," "ultimate ratios," "velocity at an instant," etc. Some of the logical problems concern whether an "ultimate ratio" is to be conceived as an actual comparison of magnitudes or as the numerical limit to an infinite sequence of such actual ratios. Some concern whether *even* given a correct mathematical understanding of Newton's "ultimate ratios" it is meaningful to attribute motion and hence velocity to a body *at* a "point of space" (*Analyst* query 30). Thus the strong claim in the *Essay on Vision* and the *Principles,* that the structure of perceptual space should be the foundation of our geometric propositions is distinctly muted, though not denied, in the *Analyst.*

John Wisdom makes the following interesting comment concerning Berkeley critique of Newton in the *Analyst* and related writings.

After all the problem that defeated Newton, the positive theory of continuity that Berkeley made no attempt to touch, is essentially the problem of Zeno, the Greek dialectician, which baffled the greatest thinkers for 2500 years.[59]

In our judgment, however, it is an exaggeration to claim that Berkeley made no attempt to "touch" the problem of "continuity" or deal with the Zenonian paradoxes. The doctrine of the sensible arithmetization of finite extensive intervals, had for Berkeley the important property of avoiding these "paradoxes." The paradox of "composition" (how a finite segment can "composed" of dimensionless points) is apparently resolved by positing that finite segments are composed of finite numbers of minima; kinematic paradoxes like the "Achilles" are comparabley solved by the denial that spatial (and temporal) intervals can be considered the sum of an actually infinite number of parts. Achilles is no longer required to occupy an infinite number of successive positions before the catches up with the tortoise. And although Berkeley offers no "positive theory of continuity" (in fact he denies that space and time are continuous) he clearly recognized that Newton's appeal to continuity through phoronomic considerations was insufficient from a logical point of view to justify a "zero increment" or "moment." The "moment," although kinematically conceived, still appeared to have the paradoxical character of being an actual magnitude less then any assignable magnitude.

Ultimate clarification of the concepts of function, variable and limit, to some extent vindicated Berkeley. The passage to the limit for a variable

[59] *Ibid.,* p. 27.

was no longer grounded in intuitions of geometrical or kinematic continuity; rather the continuity of a function at a point in the domain was made dependent on showing that the value of the function at that point was a limiting value. And the concept of "limit" (as well as the "continuum of real numbers" on which it depends) was shown to be comprehensible in a purely arithmetical way. We cannot fault Berkeley for not suggesting such a development, although his critique gave impetus to the final clarification of analysis.

Three important aspects of this "clarification" should be noted: (A) a clear distinction between mathematical and physical considerations. The "real number continuum" is purely a mathematical construction, having no necessary reference to the structure of physical space and time. From this point of view, the claim that physical and temporal intervals are continuous, that is, in a one-to-one correspondence with the real numbers, has the status of a fundamental assumption or postulate.[60] (B) The real number continuum is required in order to demonstrate the convergence of certain infinite series to limiting values.[61] By allowing mathematically formulated laws to range over all real number values (for example Galileo's law of "free fall" – $S = \frac{1}{2}gt^2$) we can compute (viewing this now merely as a mathematical procedure only) "instantaneous rates of change" of the variable "S" with respect to "t." It is in this sense that we can claim that the "density" of physical and temporal intervals (which is entailed by their "continuity") is a more fundamental "fiction" than the attribution of "instantaneous velocities" to bodies,

---

[60] Boyer, *Concepts of the Calculus*, pp. 286-287.
"Previous writers generally had defined a variable as a quantity of magnitude which is not constant, but since the time of Weierstrass it has become recognized that the ideas of variable and limit are not essentially phoronomic, but involve purely static considerations ... If for any value x_0 of the set and for any sequence of positive numbers $\delta 1$, $\delta 2$, ... δn, however small, there are in the intervals $x_0 - \delta i$, $x_0 + \delta i$ others of the set, this is called continuous ... Similarly for a continuous function ... Weierstrass defined $f(x)$ as continuous, within certain limits of x, if for an value x_0 in this interval and for an arbitrarily small positive number ϵ, it is possible to find an interval about x_0 such that for all values in this interval the difference $f(x) - f(x_0)$ is in absolute value less then ϵ ... The limit of a variable or function is similarly defined. The number L is the limit of the function $f(x)$ for $x = x_0$, given an arbitrarily small number ϵ, another number δ can be found such that for all values of x differing from x_0 by less than δ, the value of $f(x)$ will differ from that of L by less then ϵ.
[61] See Boyer, *Concepts of the Calculus, op. cit.*, p. 281, and Richard Courant and Herbert Robbins, *What is Mathematics?* (New York: Oxford University Press, 1941), pp. 289-297. For example to demonstrate that any monotone increasing sequence that has an upper bound is convergent to a limit, requires that we assume the domain of real numbers. Within the domain of rationals for example, we cannot show that the sequence of decimals, 1, 1.4, 1.41, 1.414, ... which represent the continuing approximation to the square root of two, converges or has a limit, since there is no least rational upper bound to the sequence.

where these latter are considered the "physical" correlates to the mathematically computed "limits." Thus modern *Analysis* would agree with Berkeley that the concept of the "infinitesimal" is unintelligible, but would differ on the issue of whether the "density" of spatial and temporal intervals can be dispensed with as a conceptual presupposition required in order to view the results of analysis (differentiation, for example) to be applicable to physical reality.

CONCLUSION

BERKELEY AND NEWTON

What is apparent to one who reads the Berkeleian corpus including the *Siris,* is how close to Newton's is Berkeley's conception of science. By his own account Berkeley, in his view of scientific explanation as deduction of the event (or law) from more encompassing laws, is indebted to Newtons' discussion of method in the *Principia.* The "hypothetico-deductive" method embodies what is meant by scientic explanation: deriving general rules from the "phenomena" and in a reverse procedure predicting new phenomenal consequences from these laws, allowing a continual process of confirmation. With respect to events that have occurred, to explain or "demonstrate" the "phenomenon," consists in showing that its occurrence is a logical consequence of the accepted laws.

Such "mechanical" or "causal" explanation is carefully distinguished from "metaphysical" explanation in terms of efficient causation. Although Berkeley allows causal efficacy to both divine and human wills, basing this on the intuition of volitional "power," the principle of divine causation is rooted in a more fundamental requirement, the need to make intelligible the origin, continued existence, and order of the phenomenal world. Berkeley, perhaps more than Newton, consistently carries out the Cartesian doctrine of the absolute passivity of matter and the consequent view of the existence in time of the material world as a "continuous creation."

In some important ways, Berkeley's views appear closer in spirit to those of Newton's defender, Samuel Clarke, than to Newton himself.[1] We are thinking particularly of Clarke's defense of the Newtonian position in his well known correspondence with Leibniz.[2] Clarke recognizes that given the doctrine of the absolute passivity of matter, it is no longer

[1] and [2] Alexander, H. G. ed., *The Leibniz Clarke Correspondence; op. cit.*

correct to view the laws of impact as having privileged status in the sense that all laws of motion must ultimately be reducible to them. In his Fifth Reply to Leibniz, discussing the problem of mind-body interaction, Clarke comments:

But is not God an immaterial substance? And does he act upon matter? And what greater difficulty is there in conceiving how an immaterial substance should act upon matter, then in conceiving how matter acts upon matter? Is it not as easy to conceive, how certain parts of matter may be obliged to follow the motions and affections of the soul, without corporeal contact; as that certain portions of matter should be obliged to follow each other's motions by the adhesion of parts, which no mechanism can account for . . .[3]

The action of mind on matter, then, is no more a mystery than the alleged action of matter upon matter in contact phenomena; which, from Clarke's perspective, is to say, given the passivity of matter, it is no mystery at all.

With Clarke, as with Berkeley, there is no impetus to reduce gravitational phenomena (gravity as a "manifest quality") to some "mechanism," that is, to the laws of impact. Clarke, however, does, like Newton, distinguish between gravitational *effects* (e.g., the planetary orbits), the *law* of gravity and the *cause* of gravity.

That the sun attracts the earth, through the intermediate void space; that is, that the earth and the sun gravitate towards each other, or tend (whatever be the cause of that tendency) towards each other, with a force which is in a direct proportion to their masses, or magnitudes and densities together, and in an inverse deplicate proportion of their distances; and that the space betwixt them is void, that is, has nothing in it which sensibly resists the motion of bodies passing transversely through: all this is nothing but a phenomenon, or actual matter of fact, found by experience. That this phenomenon is not produced *sans moyen*, that is without some cause capable of producing such an effect; is undoubtedly true. Philosophers therefore may search after and discover that cause if they can; be it mechanical or not mechanical.[4]

Clarke, then, following Newton, accepts the distinction between force-free motion and motion (e.g., the planetary orbits) which is said to result from the application of an impressed force. Assuming there is no efficient causality within nature, it is unclear what Clarke takes to be the ontological status of such "impressed forces." There is, in fact, some justification to Leibniz's charge that in the Newtonian system the planetary

[3] *Ibid.*, pp. 116-117.
[4] *Ibid.*, p. 118.

orbits are a continual "miracle." This, however, should not, as it appears to, disturb Clarke, for if by "miracle" we refer to the immediate intervention of God, even impact phenomena are "miracles." Changes in the momentum of bodies as a result of "impact" can no more demonstrate the independent activity of matter, than relative changes of momentum in bodies spatially separated.[5] In fact, Clarke's position might require, in the interests of consistency, that so-called "inertial" motion (non-accelerated rectilinear motion) must ultimately be explained in terms of the continued intervention of God. To endow matter with a "vis inertia" (and Newton and Clarke do so endow it), appears to violate the principle that matter is passive.[6] And if we do attribute all motion, inertial or otherwise, to the immediate intervention of God, we are still faced with the question of the ontological status of "impressed forces," those alleged to account for the non-rectilinear motions of the planets.

Berkeley, in denying altogether the existence of matter, makes more explicit the doctrine of continuous creation. Reducing material objects to collections of "sense data" or "ideas," graphically expresses the view that matter is passive. And, as with Clarke, "impact" between material particles no longer has privileged status as explanation for the changes in momentum of material bodies. In the *Siris,* for example, Berkeley appears to accept the view that laws which deal with "action at a distance," might be primitive, that is, not necessarily reducible to the laws of impact.

In light of the above comments, there is some legitimacy in the claim that Berkeley envisioned mechanics purely as a kinematics, a lawful description of motion which could dispense entirely with force terms. *De Motu* is certainly in part a criticism of the vagueness and explanatory vacuousness of force terms in physics.

Yet Berkeley, like Clarke, never satisfactorily comes to terms with the status of impressed forces in the Newtonian system. And this is connected with his apparent failure to grasp completely the significance of the "law" of inertia within that system. For example, Berkeley appeals to the "first law" as evidence for the passivity of bodies; that they have no immanent vital principle which would account for their motion. The "first law," he tells us, is evidence that uniform motion is not a state dynamically distinct from rest. From the perspective of Newtonian mechanics this is correct

---

[5] *Ibid.,* p. 110. "But indeed, all mere mechanical communications of motion, are not properly action, but mere passiveness, both in the bodies that impel, and that are impelled.

[6] *Ibid.,* p. 111. Clarke in a footnote writes: "The vis inertia of matter, is that passive force by which it always continues of itself in the state 'tis in, and never changes that state, but in proportion to a contrary power acting upon it."

with respect to impressed forces but implies nothing about the existence of immanent forces (a *vis insita*). Newton himself allows such a "force," which accounts for the resistance of bodies to changes in their motion. Moreover Berkeley appears to use the mere lawfulness of motion as evidence of the passivity of bodies. For this aim, however, uniform acceleration would have served as well as uniform velocity, and it is perhaps significant that Berkeley in his appeal to mechanics for evidence of the "passivity" of matter, does not always distinguish the two sorts of motion. The principle of inertia singles out a restricted class of motions as "force-free" in terms of impressed forces. If one accepts, as Berkeley evidently does, the principle of inertia as an empirical assertion, then "impressed forces" cannot be definitionally identified with their effects. Although *De Motu* at times suggests that "forces" in mechanics *mean* no more than their "effects," this view is at odds with the view also expressed, that Newton's laws of motion are empirical generalizations.

In addition, Berkeley makes use of the concept of impressed force to distinguish "real" from merely "relative" motion. Such a view, although making problematical Berkeley's conception of impressed forces, places him much closer to Newton than has been indicated by commentators like Karl Popper and John Myhill.[7] Newton, too, distinguished real from relative accelerations on the ground that in the former case the body moving is subject to an impressed force. Furthermore Newton considered absolute space to be the frame of reference for such "absolute" accelerations.

Berkeley rightly points out that as far as our observations go of relative change of place, we must use as a frame of reference some material system, for example, the fixed stars. This still leaves his view quite removed from the "conventionalism" of Ernst Mach; certainly more so than is indicated by the comments of Popper and Myhill. Berkeley's view is that two conditions must be met in order to claim that a body is really in motion: (1) that its relative situation change with respect to some other body, and (2) that it be the subject of an impressed force. The fixed stars, then, apparently play the role of revealing what is in fact the case, that it is Newton's bucket which is accelerating. There is no evidence that Berkeley would have accepted Mach's position that the question of which body is really moving is a meaningless one, and that strictly speaking it is only the relative acceleration of the bucket and the fixed stars which can be said to be causally antecedent to the deformation of the water's surface.

[7] Popper Karl, *op. cit.*, and Myhill, John *op. cit.*

There is, however, an unresolved difficulty in Berkele's view. Would he grant that an object could be subject to an unbalanced impressed force and at rest? That is, would he allow a choice of reference frame which would have as a consequence a situation of dynamic inequilibrium and kinematic equilibrium with respect to the same body? There is no evidence that he thinks the choice of reference frame to be "conventional" in the sense that it is dictated *only* by conside rations of simplicity. Rather, he speaks as if it is meaningful, given relative motion of (A) and (B), to denominate one of them as "really" moved; the one acting under an impressed force. Yet if this is the case, it would seem *sufficient,* in order to denominate a body moved, that it be subject to an unbalanced impressed force, and merely add that with respect to some reference frames this motion is not observable. Berkeley's view would then be quite close to Newton's contention that internal considerations were sufficient to determine whether a body is really accelerating. This again is the dynamic significance of "absolute space" as a reference frame; for as Newton himself would have admitted, "absolute space" cannot be used as an empirical frame of reference.

Commentators like Ardley,[8] Myhill,[9] and Turbayne [10] often appear to consider Berkeley's importance to consist in his separation of the legitimate body of Newtonian science from the illegitimate metaphysical additions to this work by Newton himself, or by his philosophic defenders like Locke. Following Ardley we can call this metaphysical perspective the "mathematization of nature," the view that the world described by the mathematical physicist is somehow more "real" than the world of our ordinary sensible experience.

We are not contesting this claim that Berkeley did in fact view his contribution as separating from the legitimate core of Newtonian science, certain illegitimate interpretations of that science. On the other hand, the over emphasis of this point tends to obscure the distinct differences in Berkeley's criticism of concepts like "absolute space," "force," "material substance," and "instantaneous velocity," and tends to gloss over some of the internal difficulties in this criticism. For example, in a relatively recent monograph, Turbayne writes:

Moreover, he [Berkeley] notices that may scientific entities such as force, attraction, repulsion, and of course material substance and corpuscles, most

[8] Ardley, G. W. R., *op. cit.*
[9] Myhill, *op. cit.*
[10] Turbayne, Colin: "The Influence of Berkeley's Science on His Metaphysics," *op. cit.*

of them invested by scientists as mere explaining devices quickly came to be regarded by the Philosophers as real existing things in nature, of the same order as bodies in space. The distinction between real things existing in rerum natura and phantoms ("mathematical hypotheses") which were merely entia rationes, was lost in the modern doctrine with its inadequate Philosophy of Science and false metaphysics.[11]

Turbayne has clearly oversimplified the issue. In the first place, Berkeley's claim that force terms are merely elliptical for mathematical hypotheses appeals to Newton himself as the authority, and moreover, misunderstands the latter's point. Although Newton raised questions about the nature of the "force" that produces the "manifest effects" of gravity, he never definitionally identified the "force" with these effects. Moreover, Berkeley himself apparently makes use of "impressed forces" as efficient causes to distinguish "real" from merely "relative" motion. The problem again is not the status of the Aristotelian "substantial forms," but of the Newtonian "impressed forces," those that allegedly account for the non-rectilinear motion of the planets, among other phenomena. Even with respect to the status of "immanent" (vital) principles of motion, there is an ambiguity in Berkeley's thought. He concedes that certain material configurations, e.g., human bodies, are in fact animated by an immanent principle of motion, volition. It would then seem to be an empirical question whether any particular material body's motion is caused by an immanent vital principle.

Even the problem of the existence of "material substance and corpuscles" is not as clear-cut as the passage from Turbayne might indicate. Certainly, immaterialism, strictly interpreted, would appear to rule out the existence of "insensible particles," the configurations of which were said by Locke to constitute the "real" as opposed to the "nominal" essence of bodies. Berkeley's critique, however, of the concept of material substance is more complex. In the *Principles* and the *Three Dialogues*, "material substance" is seen as that which "underlies" or is the "substratum" which "supports" the primary qualities of extension, solidity and motion. And *this* concept is criticized as unintelligible on what appear to be two different and independently sufficient grounds. (1) If the "primary qualities" like the "secondary" are demonstrably mind-dependent, it is contradictory to assert that they qualify or are the properties of an *"unthinking substance."* (2) The concept of "support" or "substratum" is itself vacuous, that is, there is nothing in principle to which the term could refer.

In addition Berkeley's critique of "corpuscularism" as it functions in

<hr>

[11] *Ibid.*, p. 478.

Locke's version of the causal theory of perception is not essentially grounded in the principles of immaterialism. Rather, Berkeley claims that (1) Locke's version of the causal theory of perception logically precludes him from saying that certain of our "ideas" "copy" certain qualities of bodies that allegedly "cause" the former; and (2) we cannot claim that "insensible particles" *cause* our sense impressions. (2) is not the traditional claim that a causal interaction between body and mind is incomprehensible, but rather follows from the general principle that except for volition there is no efficient causal principle.

In addition, any claim such as Turbayne's that Berkeley considered the "insensible particles" of corpuscular theories as mere "entia rationes," would fail to account for the numerous passages in the *Siris* where Berkeley appears to posit the existence of such particles. The passages in the *Siris* dealing with substantial scientific matters are admittedly highly speculative, and in fact lean quite heavily on Newton's speculations on the nature of light and the aether in his "queries" appended to the *Optics*. Yet we have no reason to doubt that Berkeley took them seriously. In the earlier works the possible role of "insensible particles" in scientific theory is not really discussed; Berkeley's critical attention focuses rather on how corpuscularism is alleged to function in the causal theory of perception, that is, on the theory that the "insensible particles" are causally related to our "ideas," or sense impressions.

With respect to Berkeley's influential critique of analysis in the *Analyst* and related writings, we would note also that immaterialism appears to play a minor role. His critique of the concept of "instantaneous velocity" is rather based on the claim (a) that the concept itself is incoherent, and (b) that the then current methods of differentiation violated standards of logical rigor set by the mathematicians themselves. With respect to (b), Berkeley rightly claims that Newton's attempt to surmount the logical problems inherent in the notion of the infinitely small, by substituting for the "infinitesimal," the kinematically conceived "moment," is unsuccessful. Whatever intuitive plausibilty there is, based on the perceptual continuity of motion, in conceiving a motion as just beginning ("nascent augment"), the "moment" still has the contradictory properties of being both less then any finite quantity and greater than zero. In the actual process of differentiation, the "moment" appears to function as an actual quantity, and Berkeley properly questions how, in the same demonstration, it can be posited and then ignored.

A more general claim, however, is that constitutive of the very meaning of the term "velocity" is that a given velocity be an actual comparison (ratio) of finite spatial and temporal magnitudes. From this point of view, the concept of an "instantaneous velocity" implies the mathematically unintelligible ratio $^0/_0$. The problem with Berkeley's view here, is that it tends to conflate a semantic issue concerning the "meaning" of a term with a criticism of how Newton and his followers like Jurin interpreted the derivative. Given some ambiguity in the interpretation of Newton's "ultimate ratios," the incomprehensibility of an "instantaneous velocity" might derive from the claim that it involved the ratio $^0/_0$. This is distinct from a criticism which would claim that for a body to have a velocity it must be in motion, and that motion at an instant is logically incomprehensible.

With respect to the concept of an "ultimate ratio," there are strong suggestions in Book 1 of Newton's *Principia* that an "ultimate ratio" is conceived as the numerical limit for an infinite sequence of actual ratios, ("average velocities") though not itself a member of the sequence. Moreover Newton clearly suggests that particular functions (for example "velocity") achieve their limits: implying that in a finite time an infinite number of spatial (and temporal) segments have *actually* been traversed.

The question is then raised whether, given greater clarity about "ultimate ratios" as numerical limits of an infinite sequence of actual ratios, Berkeley would have allowed the attribution of "instantaneous velocity" to a moving body. Here a negative answer does follow from Berkeley's contention that finite spatial (and probably temporal) segments are composed of finite numbers of "sensible minima."

We have suggested that Berkeley's denial that finite segments of extension can be composed of an actually infinite number of parts is motivated, in part, by a desire to overcome the "paradoxes" that allegedly follow from positing the actual infinite. On the other hand, his explicit reasons refer us to the alleged data of perceptual experience. We have argued that such data, at least with respect spatial perception (visual or tactual) are inconclusive. It does not follow without other assumptions that because finite segments of extension do not *appear* to be composed of an infinite number of parts, they must be composed of a finite number. With respect at least to perceptual space, an alternative valid response is that it contains no intrinsic metric in terms of an aggregate of minimum parts.[12]

---

[12] Berkeley perhaps has a stronger argument with respect to temporal perception; that coming into being and disappearance are essentially temporally discrete events. For an argument that temporal perception is discrete; composed of sequences of "nows"; see: Grunbaum, A., *Modern Science and Zeno's Paradoxes, op. cit.*, pp. 45-52.

With regard to the concept of "sensible minima" more than with regard to the concepts of "force" or "material substance," there is a serious question of how compatible Berkeley's views are with Newtonian science. The seriousness of the question depends on how seriously we believe Berkeley is advancing the following composite claim.

1. That finite perceptual extensive segments are composed of finite numbers of minima.

2. There is no extension other than perceptually apprehended extension. This follows from immaterialism, and is the source of Berkeley's rejection of even the "potentially infinite" subdivision of an extensive segment. Thus microscopes etc., cannot disclose anything smaller than the sensible minimum.

3. Geometry is the science of magnitude; that is, the measurement of "length," "area," etc. is the purpose of Euclidean geometry.

4. The object of Euclidean geometry is perceptually apprehended extension.

This four-part claim suggests that there is an intrinsic metric to perceptual extension based on the number of minima a segment contains; or, expressed somewhat differently, that portions of extension can be put into one-to-one correspondence only with the natural numbers (assuming the "minimum" as the unit. Assuming something larger than the minimum as the unit, fractional magnitudes would be allowable – the number of subdivisions corresponding to the number of minima composing the unit). Admittedly these suggestions are no more than programmatic (if that) and there is in Berkeley's writings, as far as we know, no discussion of how such a "metric" would be concretely used in Geometrical Optics of Mechanics. There are comments, particularly in the *Philosophic Commentaries,* to the effect that the doctrine of sensible minima does restrict the application of the Euclidean theorems; that, for example, we must reject the view that the diagonal of a square is incommensurable with the side. We can take Berkeley to mean in general that we cannot impute to bodies any metric property not in principle ascertainable by direct measurement, where the latter, given the empirical limits to actual mensuration, presumably would disclose a "minimum" which could not be further "subdivided." This, of course, would seriously limit the use of Euclidean geometry as a method of *indirect* measurement. In addition there seems to be a serious logical inconsistency in Berkeley's doctrine. If the Euclidean axioms are said to apply to certain perceptual elements, the sides of a square, for example, we cannot simultaneously hold they do not apply to certain others, in this case, the diagonal of the square. We would have to say that the sides of such a square might be called per-

ceptually "straight," but they no longer could be considered segments of
Euclidean geodesics.

Berkeley, then, even if he allowed a conceptual distinction between
"average velocity," and "instantaneous velocity," could object to predi-
cating the latter of a body in motion on the general grounds that we
cannot impute metric properties to bodies incompatible with possible
results of direct measurement. Since instantaneous velocities are not
directly ascertainable, they would be rejected. Put another way, since an
"ultimate ratio" is the mathematical limit to an actually infinite number
of "average ratios," and since there is an empirical limit to the direct
measurement of spatial and temporal intervals (and hence their ratio) no
"ultimate ratio" can be legitimately coordinated with the actual velocity
of a body in motion.

It might be objected that Berkeley would allow "instantaneous velocity" as a "mathematical fiction" comparable to "mass points" and
"attraction" or the "lines and angles" of geometers. Such fictions would
be useful in formulating laws making use of initial or impact velocities,
though such velocities could not be truly predicated of actual bodies.

Such a view is open to two fundamental objections. (1) Instantaneous
velocity would be a derivate 'fiction'; more fundamental would be the
fiction that finite spatial and temporal segments are composed of an
actually infinite number of parts. However, to consider the actual infinite,
even as a mathematical fiction, would engender those "amusing paradoxes" that Berkeley claims to have avoided in his doctrine of sensible
minima. (2) It is not clear (contra Warnock and Ardley) that Berkeley
makes a distinction between "pure" and "applied" geometry. The passages adduced by these authors as evidence for this claim are not un-
amibguous. Moreover, from the *Philosophic Commentaries* to the
"Queries" appended to the *Analyst*, Berkeley contends that perceptual
extension is the object of geometry. We might take this to mean that the
referents for the fundamental geometric terms like "point," "line," etc.
are properties of perceptual objects. If the postulates and theorems of
geometry articulate the properties of perceptually apprehended extension,
and the latter is composed of "sensible minima," then it is unclear in
what sense Berkeley would allow "instantaneous velocity" as a "mathematical fiction."

If the "denseness" of physical space and time is rejected, (even as a
"useful" fiction) then Berkeley's views do more than separate the Newtonian wheat from the methaphysical chaff; they are in fact incompatible
with Newtonian science which requires the assumption of "denseness"
for finite spatial and temporal segments.

Berkeley does attempt, as the *Analyst* demonstrates, to show that finite considerations are sufficient to gain the results of analysis. We have looked at the proposed method of the "compensation of errors," and concluded that Berkeley does not establish it as a mathematically valid substitute for Newton's method of fluxions. Berkeley also suggests that since "we measure only assignable extensions," that "unlimited approximations complete answer the intention of geometry." (*Analyst* Query 53) Again, it is difficult to see how this suggestion can be broadened into an alternative method for solving the problem of "tangents" (differentiation). A directly measured sequence of "average velocities" would not disclose that limiting value called the "instantaneous velocity." Computation of the derivate presupposes the existence of "continuous functions" that relate, for example, distance and time; the very type of function apparently ruled out by the doctrine of sensible minima.

At times Berkeley suggests that the classical method of "exhaustion" would be a satisfactory substitute for the method of "fluxions." However, as we have suggested, the method, although denying the actual passage *to* a limit of an infinite series of states, makes use of the concept of a *potentially* infinite subdivision of a finite extensive segment, a concept incompatible with the view that a non-infinitesimal "minimum" constitutes the limit to any such division. And although Berkeley appeals to Newton's discussion in Book 1 of the *Principia* to support his view, that discussion clearly suggests that in a finite time the passage to the limit of an infinite number of states is actually accomplished. That is, although the "instantaneous velocity" is not a member of the sequence of average velocities for which it is the limit, it is "reached" by the moving body. Put in terms of the Zenonian "paradoxes," Achilles in a finite time not only traverses an infinite sequence of segments, but actually catches up with the tortoise. Thus the actually infinite subdivision of an extensive segment is required, as opposed to the "potentially" infinite subdivision in the classical method of exhaustion.

Finally we would conclude that there is in Berkeley's critique of specific concepts in mechanics and analysis, no architectonic principle under which these various criticisms can be subsumed, although they all raise important questions about the role of such concepts in Newtonian science. Certainly the erroneous claim that for every "noun substantive" there must be a sensible referent is not the significant source of the demand to posit the existence of "entities" as diverse as "general triangles," "material substance," "impressed forces," "absolute space" and "instantaneous velocities." Nor is there any univocal sense of "abstraction," such that we can say with respect to the 'entities' mentioned above, that what is being posited illegitimately is the existence of "abstract" entities. To take one example, although generic space or extension as an abstraction may have something in common with triangularity ("general triangle") as an abstraction, Berkeley's critique of "absolute space" in *De Motu* is not primarily directed against positing the existence of such "generic" space, but rather against positing "space" as an entity or thing separable from the things and processes it allegedly *contains*. And the relation of container to contained is distinctly different from the relation of genus to species. And whether ultimately it is legitimate to posit the existence of "forces" (either the Aristotelian "substantial forms" or the Newtonian "impressed forces") and "instantaneous velocities," it is difficult to see in what sense we can speak of these concepts as 'abstractions." The nominalist-conceptualist controversy has no relevance here, since Berkeley's point is that there are no *individual* examples of either "force" or "instantaneous velocity."

BIBLIOGRAPHY

BERKELEY'S WORKS

Berkeley, George *A Treatise Concerning the Principles of Human Knowledge; Three Dialogues Between Hylas and Philonous; Correspondence with Samuel Johnson.* Ed., Colin Murray Turbayne; New York: Bobbs Merrill Co., Inc. (Library of Liberal Arts) 1965.
An Essay Towards a New Theory of Vision; The Theory of Vision or Visual Language Vindicated and Explained. Ed., Colin Murray Turbayne. New York: Bobbs Merrill Co., Inc. (Library of Liberal Arts) 1963.
De Motu; The Analyst; A Defense of Free Thinking in Mathematics; Reasons for not Replying to Mr. Walton's Full Answer; of Infinites. Ed., A. A. Luce (including Luce's English translation of *De Motu*) in *The Works of George Berkeley, Bishop of Cloyne,* Ed. A. A. Luce and T. E. Jessop. Thomas Nelson and Sons, London, 1951. Vol. 4.
The Siris, Ed., T. E. Jessop, in Works, Vol. 5.
Philosophical Commentaries. Ed., A. A. Luce in *Works,* Vol. 1.
Alciphron, or The Minute Philosopher. Ed., T. E. Jessop in *Works,* Vol. 3.

WORKS ON BERKELEY – BOOKS

Abbott, Thomas K. *Sight and Touch.* London: Longmans Green, 1864.
Ardley, Gavin *Berkeley's Renovation of Philosophy.* The Hague, Martinus Nijhoff, 1968.
Armstrong, David M. *Berkeley's Theory of Vision.* Melbourne; Melbourne University Press, 1961.
Bailey, Samuel *A Review of Berkeley's Theory of Vision.* London, 1842.
Bracken, H. M. *The Early Reception of Berkeley's Immaterialism.* Doctoral dissertation, State University Iowa, 1956.
Givner, David *A Study of George Berkeley's Theory of Linguistic Meaning.* Doctoral Dissertation Columbia University, 1959.
Hicks, G. Dawes *Berkeley,* London: 1932.
Jessop, T. E. *George Berkeley.* London: Longman's Green, 1959.
Johnston, G. A. *The Development of Berkeley's Philosophy.* New York: Russell and Russell Inc., 1965.
Luce, A. A. *Berkeley and Malbranche.* New York: Oxford, 1934.
Warnock, G. J. *Berkeley.* Baltimore: Penguin Books, 1969 (originally published in 1953).
Wild, John *George Berkeley.* (New York: Russell and Russell Inc., 1962; first published – 1936 – Harvard University Press.
Wisdom, John *The Unconscious Origins of Berkeley's Philosophy.* New York, Hillary House, 1957.

WORKS ON BERKELEY – ARTICLES

Ardley, Gavin "Berkeley's Philosophy of Nature," Bulletin No. 63, Series No. 3, University of Aukland (1962).

Bracken, H. M. "Substance in Berkeley," *From New Studies in Berkeley's Philosophy*, ed. Warren E. Steinkraus, New York: Holt, Rinehart, and Winston, 1966.

Cajori, F. "Discussion of Fluxions: From Berkeley to Woodhouse," *American Mathematical Monthly*, Vol. XXII, No. 5 (May 1915) 143-149.

Crombie, A. C. "A Note on the Descriptive Conception of Motion in the 14th Century," *British Journal of Philosophy* 4, 13 (May 1953) 46-51.

Davis, John W. "The Molyneux Problem" Journal History of Ideas 21, 3 (July-Sept. 1960) 392-408.

"Berkeley Newton and Space," in R. Butts and J. W. Davis eds., *The Methodological Heritage of Newton* (London: Basil Blackwell, 1970) 57-74.

Furlong, E. J. "Berkeley's Theory of Meaning," *Mind* 73 (1964) 437-438.

Hinrich, G. "The Logical Positivism of *De Motu*," *Review of Metaphysics* III (1950) 491-505.

"Berkeley on Size and a Common World," *Personalist* (1951) 251-258.

Jessop, T. E. "Berkeley and the Contemporary Physics," *Review International Philosophy,* 7 (1953) 87-100.

Johnston, G. A. "The Influence of Mathematical Conceptions on Berkeley's Philosophy," *Mind* 25 (1961) 177-192.

Martin, C. B. and Armstrong, D. M. eds., *Locke and Berkeley*, A Collection of Critical Essays, (New York: Doubleday and Co., Inc.) 1968.

Myhill, John "Berkeley's *De Motu*," in *George Berkeley* (eds., Pepper Stephen C., Aschenbrenner, Karl and Mates, Benson) University of California Press, (1957) 141-157.

Pirenne, M. H. "Physiological Mechanisms in the Perception of Sight and Berkeley's Theory of Vision," *British Journal Philosophy of Science* 4, (1953-54) 13-21.

Strong, Edward W. "Mathematical Reasoning and its Objects," in *George Berkeley* (eds., Pepper, Aschenbrenner and Bates) University of California Press, (1957) 20-36.

Titley, G. W. I. "Berkeley and Helmholtz – Theories on Space Perception," *The Optometric Weekly,* 46 (1955) 1823-1826.

Turbayne, C. M. "The Influence of Berkeley's Science on His Metaphysics," *Philosophy and Phenomenological Research* No. 4, Vol. XVI (June 1956) 476-487.

"Berkeley and Ronchi on Optics," *Proceedings 12th International Congress of Philosophy* 12 (1961) 453-460.

"Berkeley and Molyneux on Retinal Images, *Journal History of Ideas,* 17 (1956) 127-129.

White, Allen R. "A Linguistic Approach to Berkeley's Philosophy," *Philosophy and Phenomenological Research,* Vol. XVI, No. 2 (1955) 172-187.

Whitrow, G. J. "Berkeley's Philosophy of Motion," *Britisch Journal for Philosophy of Science* 4, 13 (1953-54) 37-45.

Wisdom, John "The Analyst Controversy-Berkeley's Influence on the Development of Mathematics," *Hermathena*, No. 29 (1939) 3-29.

"The Analyst Controversy-Berkeley as a Mathematician," *Hermathena,* No. 53.

"The Compensation of Errors in the Method of Fuxions," *Hermathena,* No. 57 (1941) 49-81.

GENERAL WORKS – BOOKS

Alston, William P. *Philosophy of Language,* Englewood Cliffs: Prentice Hall, 1964.

Alexander, H. G. ed., *Leibniz Clarke Correspondence,* Manchester: Manchester University Press, 1956.

Beracerraf, Paul and Putnam, Hillary eds., *Philosophy of Mathematics,* Englewood Cliffs: Prentice Hall, 1964.

Boyer, Carl B. *The History of the Calculus and its Conceptual Development,* Hafner Pub. Co., 1949-reprinted New York: Dover, 1959.

History of Mathematics, New York: John Wiley and Sons, Inc., 1968.

Buchdahl, Gerd *Metaphysics and the Philosophy of Science, The Classical Origins: Descartes to Kant,* Cambridge, MIT Press, 1969.

Chiu, H. Y. and Hoffman, W. F. eds., *Gravitation and Relativity,* New York: Benjamin, 1964.

Colodny, Robert ed. The Edge of Certainty: *Essays in Contemporary Science and Philosophy,* Englewood Cliffs: Prentice Hall 1965.

Courant, Richard and Robbins, Herbert *What is Mathematics,* New York: Oxford University Press, 1941.

Danto, A. and Morganbesser, S. eds., *Philosophy of Science,* New York: Meridian Books, 1960.

Farber, Marvin *Foundations of Phenomenology,* Albany: State University of New York Press, 1943.

Frege, Gottlob *The Foundation of Arithmetic,* trans. J. L. Austin, New York: Harper and Bros., 1950.

Galilei, Galileo *Dialogues Concerning Two New Sciences,* TRANS. Henry Crew and Alfonsio De Salvio (first pub. 1914) New York: McGraw Hill, 1963.

Grunbaum, Adolph *Modern Science and Zeno's Paradoxes,* Middletown: Wesleyan University Press, 1967.

Geometry and Chronometry in Philosophical Perspective, Minneaspolis: University of Minneapolis Press, 1962.

Hanson, Norwood Russel *Patterns of Discovery,* Cambridge: Cambridge University Press, 1961.

Hall, Robert and Hall, Marie Boaz *Unpublished Papers of Isaac Newton,* Cambridge: Cambridge University Press, 1962.

Hume, David *Enquiries Concerning the Human Understanding.* (Reprinted from the 1777 edition, ed., by L. A. Selby-Bigge) London: Oxford University Press, 1966.

Treatise, Vol. 1, London: Dent (Everyman's Library), 1911 (originally pub. in 1738).

James, William *The Principles of Psychology,* Vol. II, (original copyright 1890), New York: Dover (reprint) 1950.

Jammer, Max *Concept of Force,* Cambridge: Harvard University Press, 1957.

Concept of Space, Cambridge, Harvard University Press, 1954 (Reprinted Harper Torchbook, 1960).

Koyre, Alexander *From the Closed World to the Infinite Universe,* Baltimore: John Hopkins's Press, 1957 (Reprinted Harper Torchbook 1958).

Newtonian Studies, Chicago: University of Chicago Press, 1968.

Leibniz, G. W. F. *Selections,* ed., Phillip P. Wiener, New York: Charles Scribners and Sons, 1951.

Locke, John *Essay Concerning Human Understanding,* Chicago: Henry Regnery and Co., 1956.

Mach, Ernst *The Science of Mechanics* sixth ed. (first German ed. 1883) trans. Thomas J. McCormack, Lasalle I 11.: Open Court Publishing Co., 1960.

Kuhn, Thomas *The Structure of Scientific Revolutions,* Chicago, University of Chicago Press, 1962.

Korner, Stephen *The Philosophy of Mathematics,* London: Hutchinson and Co., Ltd., 1960. (Reprinted as a Harper Torchbook 1962).

Morganbesser, S., Suppes, P., and White eds., *Essays in Honor of* Ernest Nagel, St. Martin's Press, 1969.

Nagel, Ernest *The Structure of Science,* New York: Harcourt Brace and World, Inc., 1961.

Newton, Isaac *Mathematical Principles of Natural Philosophy* (3rd ed.) trans. Andrew Motte; revised by Florian Cajori, Chicago: Encyclopedia, Britannica (Great Books of the Western World) 34, 1952.

Optics, New York: Dover Pub. Co., 1952 (based on the forth edition, London, 1730).

Quine, W. O. *Word and Object,* New York: John Wily and Sons, 1960.

Rescher, Nicholas *The Philosophy of Leibniz,* Englewood Cliffs: Prentice Hall, 1967.

Toulman, Stephen *Philosophy of Science,* Hutchinson and Co., Hutchinson University Library, (London) 1953 (reprinted Harper Torchbook 1960).

GENERAL WORKS – ARTICLES

Ballard, K. E. "Leibniz's Theory of Space and Time," *Journal History of Ideas,* Vol. 21 (1960) 49-65.

Cajori, Florian "History of Zeno's Arguments on Motion," *American Mathematical Monthly,* XXII (1915) 1-6, 39-47, 77-82, 109-115, 143-49, 179-86, 215-20, 253-58, 292-97.

"The Purpose of Zeno's arguments on Motion," *Isis* III (1920) 7-20.

Gay, John H. "Matter and Freedom in the Thought of Samuel Clarke," *Journal History of Ideas,* Vol. XXIV, No. 1 (Jan.-March 1963) 85-103.

Hesse, Mary B. "Action at a Distance in Classical Physics," *Isis,* Vol. 46, No. 146 (Dec. 1955) 336-354.

Lewis, Douglas "Some Problems of Perception," *Philosophy of Science,* Vol. 37, No. 1 (March 1970).

Suchting, W. A. and Hunt, I. E. "Force and Natural Motion," *Philosophy of Science,* Vol. 36. No. 3 (Sept. 1969) 233-250.